Simple Atomic and Molecular Systems

New results and applications

Simple Atomic and Molecular Systems

New results and applications

Eugene Oks
Physics Department, Auburn University, Auburn AL, USA

IOP Publishing, Bristol, UK

ISBN 978-0-7503-3679-6 (ebook)
ISBN 978-0-7503-3677-2 (print)
ISBN 978-0-7503-3680-2 (myPrint)
ISBN 978-0-7503-3678-9 (mobi)

DOI 10.1088/978-0-7503-3679-6

Version: 20210201

IOP ebooks

British Library Cataloguing-in-Publication Data: A catalogue record for this book is available from the British Library.

Published by IOP Publishing, wholly owned by The Institute of Physics, London

IOP Publishing, Temple Circus, Temple Way, Bristol, BS1 6HG, UK

US Office: IOP Publishing, Inc., 190 North Independence Mall West, Suite 601, Philadelphia, PA 19106, USA

To my daughter Alice with love

Contents

Author biography

Eugene Oks

Eugene Oks received his PhD degree from the Moscow Institute of Physics and Technology and later the highest possible degree of Doctor of Sciences from the Institute of General Physics of the Academy of Sciences of the USSR by the decision of the Scientific Council led by the Nobel Prize winner, academician A M Prokhorov. According to the Statute of the Doctor of Sciences degree, this highest degree is awarded only to the most outstanding PhD scientists who founded a new research field of great interest. Oks worked in Moscow (USSR) as the head of a research unit at the Center for Studying Surfaces and Vacuum, then at the Ruhr University in Bochum (Germany) as an invited professor, and for the last 30 years at the Physics Department of Auburn University (AL, USA) in the role of Professor. He has conducted research in five areas: atomic and molecular physics, astrophysics, plasma physics, laser physics, and nonlinear dynamics. He founded/co-founded and developed new research fields, such as intra-Stark spectroscopy (a new class of nonlinear optical phenomena in plasmas), masing without inversion (advanced schemes for generating/amplifying coherent microwave radiation), and quantum chaos (nonlinear dynamics in the microscopic world). He also developed a large number of advanced spectroscopic methods for diagnosing various laboratory and astrophysical plasmas—methods that were subsequently used and continue to be used by many experimental groups around the world. He recently revealed that there are two flavors of hydrogen atoms, as proven by the analysis of atomic experiments; there is also a possible astrophysical proof from observations of the 21 cm radioline from the early Universe. He showed that dark matter or at least part of it can be represented by the second flavor of hydrogen atoms. He has published about 500 papers and eight books, including the books *Plasma Spectroscopy: The Influence of Microwave and Laser Fields*, *Stark Broadening of Hydrogen and Hydrogenlike Spectral Lines in Plasmas: The Physical Insight*, *Breaking Paradigms in Atomic and Molecular Physics*, *Diagnostics of Laboratory and Astrophysical Plasmas Using Spectral Lineshapes of One-, Two-, and Three-Electron Systems*, *Unexpected Similarities of the Universe with Atomic and Molecular Systems: What a Beautiful World*, *Analytical Advances in Quantum and Celestial Mechanics: Separating Rapid and Slow Subsystems*, and *Advances in X-Ray Spectroscopy of Laser Plasmas*. He is the Chief Editor of the journal *International Review of Atomic and Molecular Physics*. He is a member of the editorial boards of five other journals: *Symmetry*, *American Journal of Astronomy and Astrophysics*, *Dynamics*, *Open Journal of Microphysics*, and *Open Physics*. He is also a member of the International Program Committees of two series of conferences: Spectral Line Shapes, as well as the Zvenigorod Conference on Plasma Physics and Controlled Fusion.

IOP Publishing

Simple Atomic and Molecular Systems
New results and applications
Eugene Oks

Chapter 1

Introduction

Simplicity is the ultimate sophistication.

—Leonardo da Vinci

Nature is pleased with simplicity. And nature is no dummy.

—Isaac Newton

Make everything as simple as possible, but not simpler.

—Albert Einstein

Simplicity is the keynote of all true elegance.

—Coco Chanel

This book is devoted to simple atomic and molecular systems—specifically to systems consisting of no more than three particles. Simple atomic and molecular systems are the test-bench of atomic and molecular physics in general and of quantum mechanics in particular. This is because simple atomic and molecular systems may allow analytical solutions, not only as an isolated system but also in certain environments—the solutions revealing a rich physics despite the systems being simple. In addition, simple atomic and molecular systems and the phenomena involving them sometimes have unexpected similarities with some astrophysical objects and phenomena—see [1].

Theoretical studies of simple atomic and molecular systems began about a century ago. One might thus think that there is only old news about these simple systems. However, in reality even now theorists are finding new twists and unexpected, counterintuitive results concerning simple atomic and molecular systems in various situations. Presenting such results is the focus of this book.

One more thing—analytical solutions concerning simple atomic and molecular systems in various situations reveal their beauty and elegance. These aesthetic concepts play a very important role in theoretical physics (in distinction to computational physics). They are intimately connected to the physical insight provided by analytical solutions. Thus the above epigraph quoting Coco Chanel rings true—even if she did not intend it to relate to theoretical physics.

doi:10.1088/978-0-7503-3679-6ch1

The book is structured as follows. Chapter 2 presents a fundamental discovery proving that hydrogen atoms come in more than one 'flavor' (similarly to quarks). This breakthrough was achieved using the standard quantum mechanics and without changing the physical laws. The existence of the second (unusual) flavor of hydrogen atoms has proof from atomic experiments. In addition, there is also a possible astrophysical proof. We show that dark matter, or part of it, is possibly represented by the second flavor of hydrogen atoms. This explanation seems to be more natural than other hypotheses, all of which resort to largely unspecified, experimentally never-discovered subatomic particles or to significantly changing the physical laws. The results presented in this chapter are based on [2–4].

Chapter 3 presents the monopole contribution to the width of hydrogen-like spectral lines in plasmas. A counterintuitive result is discussed in this chapter. The results presented in this chapter are based on [5].

Chapter 4 presents the consequences of the separation or non-separation of the center-of-mass motion from the relative motion for systems, such as hydrogen atoms subjected to a uniform magnetic field or to an electric field, the latter being either uniform or non-uniform. The application of this provides new results for the dynamical Stark broadening of hydrogen spectral lines by plasma ions. The results presented in this chapter are based on [6, 7].

Chapter 5 presents analytical results for the ionization rate of atoms by a laser field in the regime of tunneling ionization. These analytical results are valid for an arbitrary strength of the laser field. The results presented in this chapter are based on [8].

Chapter 6 presents the application of Dirac's generalized Hamiltonian dynamics to a charged spherical harmonic oscillator. The results presented in this chapter are based on [9].

Chapter 7 presents selected applications for spectroscopic diagnostics of plasmas. Some results presented in this chapter are based on [10].

Chapter 8 presents applications for enhancing the gain of plasma-based x-ray lasers. The results presented in this chapter are based on [11, 12].

Chapter 9 presents applications for designing advanced generators of coherent infra-red or microwave radiation (masers). The results presented in this chapter are based on [13, 14].

Chapter 10 presents a scenario providing a possible explanation of the entire history of the expansion of the Universe—both the era of decelerating expansion and the current era of accelerated expansion. In this model there is no need to resort to dark energy or to new gravitational degrees of freedom (mass gravity etc), in distinction to all other hypotheses.

Some additional results are presented in appendices A, B, C, and D.

References

[1] Oks E 2019 *Unexpected Similarities of the Universe with Atomic and Molecular Systems: What a Beautiful World* (Bristol: IOP Publishing)

[2] Oks E 2001 *J. Phys.* **B 34** 2235

[3] Oks E 2020 *Atoms* **8** 33
[4] Oks E 2020 *Res. Astron. Astrophys.* **20** 109
[5] Oks E 2020 *Plasma* **3** 180
[6] Oks E 2017 *Int. Rev. Atomic Mol. Phys.* **8** 41
[7] Oks E 2018 *J. Phys. Communications* **2** 045005
[8] Gavrilenko V P, Oks E and Canad 2011 *J. Phys.* **89** 849
[9] Oks E 2020 *Symmetry* **12** 1130
[10] Gavrilenko V P and Oks E 2006 *AIP Conf. Proc.* **874** 242
[11] Oks E 2000 *J. Phys. B: At. Mol. Opt. Phys.* **33** L801
[12] Gavrilenko V P and Oks E 2004 *Eur. Phys. J.* D **28** 253
[13] Gavrilenko V P and Oks E 1995 *J. Phys. B: At. Mol. Opt. Phys.* **28** 1433
[14] Gavrilenko V P and Oks E 1995 *Phys. Rev. Lett.* **74** 3796

Chapter 2

Two flavors of hydrogen atoms: a possible explanation of dark matter

Not from pictures all being drawn,
And not sewn for show buzz,
The invisibles live around,
All around—around us.
They're upset by not being visible,
By offenses all they took,
But sometimes all these invisibles,
May, in fact, have a decent look.

—Russian student song translated by the author of this book. Below is the original in Russian.

Не срисованы с картинки,
И не сшиты на показ,
Проживают невидимки,
Невидимки среди нас.
Их не видно,им обидно,
Пуще видимых обид,
Но порой у невидимок,
Может быть приличный вид.

2.1 The mystery and its resolution through a fundamental discovery

Hydrogen atoms are the test-bench of atomic physics in general and of quantum mechanics in particular. Despite their simplicity and despite the fact that quantum solutions for hydrogen atoms were obtained almost a century ago, there remained a

doi:10.1088/978-0-7503-3679-6ch2

long-standing mystery (for three dozen years) of the striking discrepancy between the theoretical and experimental results for the high-energy tail of the linear momentum distribution in the ground state of hydrogen atoms (GSHA). The eventual resolution of this mystery and the subsequent theoretical developments that led, in particular, to far reaching astrophysical consequences and a quite natural explanation of dark matter, are the subject of this chapter. In this chapter we present results published in [1, 2] that are further developments of the results from [3] (the results of [3] being later reproduced in chapter 2 of [4]). For brevity, below we use the word 'momentum' to mean 'linear momentum'.

In 1935 Fock [5] derived the following non-relativistic result, the momentum distribution $dw = f_{quant}(p)dp$ in the GSHA:

$$f_{quant}(p) = 32p^2p_0^5\Big/\left[\pi\left(p_0^2 + p^2\right)^4\right], \quad p_0 \equiv me^2/\hbar. \tag{2.1}$$

Here, $p_0/Z \approx 1.992 \times 10^{-19}$ g·(cm s^{-1}) corresponds to the atomic unit of the linear momentum and m is the reduced mass of the electron. For the high-energy tail of the momentum distribution (HTMD), equation (2.1) yields

$$f_{quant}^{As}(p) \equiv f_{quant}(p \gg p_0) \propto 1/p^6. \tag{2.2}$$

However, in [6, 7] it was demonstrated that the experimental HTMD has a dramatically different behavior: $\sim 1/p^u$, where the exponent u is *at least 1.5 times smaller than in the quantum HTMD*. This puzzling discrepancy is amplified by the fact that there is a significant range of momenta where the above non-relativistic quantum result should be legitimate.

Figure 2.1 shows the ratio of the theoretical HTMD (from equation (2.2)) to the experimental HTMD [6, 7]. It is seen that this ratio, i.e. the relative discrepancy between the experimental and theoretical results, can be many orders of magnitude.

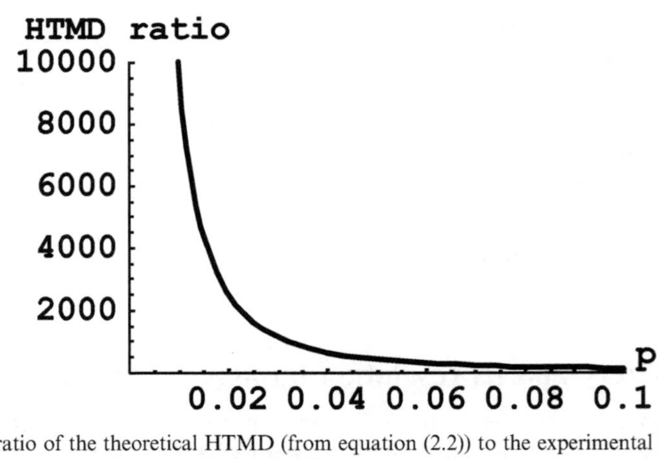

Figure 2.1. The ratio of the theoretical HTMD (from equation (2.2)) to the experimental HTMD [6, 7]. The linear momentum p is in units of mc, where c is the speed of light.

The following questions remained:

1. Would the relativistic treatment of this problem using the Dirac equation eliminate this huge discrepancy?
2. Would an allowance for the finite size of the nucleus (proton) eliminate this huge discrepancy?

The answer to the first question was negative. No wonder—the huge discrepancy was already in the non-relativistic range of momenta.

As for the second question, it can be reformulated as follows: is the singularity at the origin—in models considering the point-like nucleus—the cause of the above huge discrepancy? The question about the singularity brings us to the problem of *singular solutions* of the Schrödinger or Dirac equations. This is a story inside a story. It is *one of the most fundamental problems* in quantum mechanics.

As a matter of fact, both the Schrödinger and Dirac equations for hydrogen atoms yield two formal solutions. One solution is regular at the origin, while another solution is singular at the origin. Below are some details.

The wave functions in the momentum representation $P_{nl}(p)$ and in the radial wave function in the coordinate representation $R_{nl}(r)$ for potentials having spherical symmetry are related to each other as follows—see e.g. [8] (here and below, n is the principal quantum number and l is the angular momentum quantum number):

$$P_{nl}(p) = [r/p]^{1/2}(i^{-l}/\hbar)\int_0^\infty \mathrm{d}r\, J_{l+1/2}(pr/\hbar)R_{nl}(r)r, \tag{2.3}$$

where $J_{l+1/2}(z)$ is the Bessel function.

For the GSHA, from the Schrödinger equation, we obtain the following expression for the distribution function (2.1):

$$f_{\text{quant}}(p) = |P_{10}(p)|^2 P^2, \quad P_{10}(p) = [2/(\pi)]^{1/2}(1/p)\int_0^\infty \mathrm{d}r\, \sin(pr/\hbar)R_{10}(r)r. \tag{2.4}$$

From equation (2.4) it is clear that any problem with $P_{10}(p)$ at large p means that there is a problem with $R_{10}(r)$ at small r. The textbook (regular) solution of the Schrödinger equation at small r yields (see e.g. [8, 9]) $R_{nl}(r) \propto r^l$, meaning that $R_{10}(r) \approx$ const at small r. The fact that the experimental HTMD decreases much more slowly than its theoretical non-relativistic counterpart indicates that the actual function $R_{10}^{\text{true}}(r)$ should have a relatively strong singularity at small r: $R_{10}^{\text{true}}(r) \propto 1/r^q$, $q \geqslant 1$.

The formal singular solution yields (see e.g. [8, 9]) $R_{nl}(r) \propto 1/r^{l+1}$. For $l \geqslant 1$, the singular solution cannot be normalized (the corresponding integral $\int_0^\infty |R_{nl}(r)|^2 r^2 \mathrm{d}r$ diverges at $r = 0$).

As for $l = 0$, there is no problem with the normalization integral at $r = 0$. Most quantum mechanics textbooks reject the singular solution of the Schrödinger equation without an explanation, while the explanations in [8, 10] were incorrect, as shown in [3]. Oks [3] derived the second (singular) solution of the Schrödinger equation for the GSHA in the explicit form (for the first time):

$$R_{10}(r) = \text{const}\{E_i(2r) - [\exp(2r)]/(2r)\}\exp(-r), \tag{2.5}$$

where $E_i(z)$ is the exponential integral function. Here and below, we utilize the Coulomb units $Z = \hbar = m = e = 1$, except for expressions where Z, or $\hbar$, or m, or e is shown explicitly. At small r, this solution reduces to

$$R_{10}(r) \approx \text{const}[1/r + \log(1/r^2)]. \tag{2.6}$$

The functions on the right sides of equations (2.5) and (2.6) are non-analytic functions at $r = 0$, but the corresponding normalization integral converges at $r = 0$. Thus the rejection of this solution in all textbooks on the grounds of its 'bad' behavior at $r = 0$ was without merit. In [3] it was shown that this singular (at $r = 0$) solution has to be rejected for a different reason: the normalization integral diverges at $r \to \infty$ (because the solution essentially becomes $\exp(2r)$ at $r \to \infty$).

The Dirac equation for hydrogenic atoms/ions of nuclear charge Z, for the radial part $R_{n'j}(r)$ of the coordinate wave function at small r, yields the following two options (see e.g. [11–14]):

$$R_{n'j}(r) \propto 1/r^{1+s}, \quad s = \pm[(j + ½)^2 - (Z\alpha)^2]^{1/2}, \quad \alpha \equiv e^2/(\hbar c) \approx 1/137. \tag{2.7}$$

Here j is the quantum number of the total angular momentum and n' is the radial quantum number. For the GSHA, one obtains $n' = 0$ and $j = 1/2$. Consequently, equation (2.7) reduces to $R_{0½}(r) \propto 1/r^s$, $s = 1 \pm [1 - (Z\alpha)^2]^{1/2}$.

Our focus is on the situation where $Z\alpha \ll 1$. In this situation the solution dealing with $s = 1 - [1 - (Z\alpha)^2]^{1/2} \approx (Z\alpha)^2/2 \approx 2.66 \times 10^{-5}Z^2$ is considered legitimate (its weak singularity at $r = 0$ does not prevent the normalization (let us call it 'regular')). The formal solution corresponding to $s = 1 + [1 - (Z\alpha)^2]^{1/2} \approx 2$ is considered illegitimate: it is strongly singular at $r = 0$ and cannot be normalized.

The above solutions of the Schrödinger and Dirac equations were obtained for the model where the nucleus is considered point-like. However, in reality, the nucleus occupies some volume. For a finite nuclear size, relativistic studies [11, 15–19], performed previously [3], concluded that singular solutions of the Dirac equation are illegitimate for the Coulomb potential with $Z < 1/\alpha$, including hydrogen atoms.

In [3] the author studied the following general problem. Under his consideration was an *arbitrary* spherically symmetric interaction potential $V(r)$, having one form in the interior region ($r < R$) of the nucleus of radius R and another form in the exterior region ($r > R$). According to previous studies [11, 15–19], for the interior and exterior solutions of the Dirac equation to match at the boundary (at $r = R$) the ratio $f(r)/g(r) \equiv \rho(r)$ of the two radial components of the Dirac bispinor should not have any discontinuity at $r = R$. Thus, in this general problem, [3] started directly with the equation for $\rho(r)$ (which is a Ricatti-type equation) for the ground state:

$$\rho' = -2\rho/r + [V(r) - E + 1] + [V(r) - E - 1]\rho^2. \tag{2.8}$$

Here, $\rho' \equiv d\rho/dr$ and E is the total energy; the natural units $c = \hbar = m = 1$ are used. After denoting $u(r) \equiv \rho(r)r^2$, equation (2.8) reduces to

$$u' = r^{-2}[V(r) - E - 1][u(r)]^2 + [V(r) - E + 1]r^2. \tag{2.9}$$

In [3] the author analyzed equations (2.8) and (2.9) at $r \ll 1$ for the relevant situation where $R \ll 1$. For a *regular* solution of the interior region ($r < R$), it is expected that $\rho_{reg}(r) \ll 1$, consequently $u_{reg}(r)/r^2 \ll 1$. For this reason, the first term on the right-hand side of equation (2.9) can be neglected. This leads to

$$u_{reg}(r) \approx \int_0^r V(r')r'^2 dr' + (1 - E)r^3/3. \tag{2.10}$$

For a *singular* solution in the exterior region ($R < r \ll 1$), it is expected that $\rho_{sing}(r) \gg 1$—consequently $u_{sing}(r)/r^2 \gg 1$. For this reason, the second term on the right-hand side of equation (2.9) can be neglected. This leads to

$$du_{sing}/u_{sing}^2 \approx [V(r) - E - 1]/r^2, \tag{2.11}$$

so that

$$u_{sing}(r) \approx \left\{ \int_r^\infty [V(r')/r'^2] Dr' - (1 + E)/r \right\}^{-1}. \tag{2.12}$$

Then the matching condition $u_{reg}(R) = u_{sing}(R)$ can be represented as follows:

$$\int_0^R V(r')r'^2 dr' + (1 - E)r^3/3 \approx \left\{ \int_R^\infty [V(r')/r'^2] dr' - (1 + E)/r \right\}^{-1}. \tag{2.13}$$

Then, the author analyzed [3] whether the matching condition (2.13) is satisfied for the GSHA. The charge density inside protons has a maximum at $r = 0$—this is known from experiments on the elastic scattering of electrons on protons [20–22]. For the potential, corresponding to such a charge density inside the nucleus, the matching condition (2.13) is satisfied. This means that for the GSHA, the regular solution of the Dirac equation inside the proton can be matched with the *singular* solution of the Dirac equation outside the proton.

Consequently, for $r \geqslant R$ the radial part of the Dirac bispinor for the GSHA should be a linear combination of the corresponding regular and singular solutions. Finally, the author obtained the following for the GSHA [3]:

$$\begin{aligned} f(r) &\approx -\left\{1/r^{1-\gamma} + [(Z\alpha)^2(2\lambda)^{2\gamma}]^{-1} 2\Delta / r^{1+\gamma}\right\}(2\lambda)^{1/2+\gamma}(1 - E)^{1/2}\exp(-\lambda r), \\ g(r) &\approx \left\{1/r^{1-\gamma} + [(2\gamma - 1)(Z\alpha)^2(2\lambda)^{2\gamma}]^{-1} \Delta / r^{\gamma}\right\}(2\lambda)^{1/2+\gamma}(1 + E)^{1/2}\exp(-\lambda r), \\ \gamma &\equiv (1 - Z^2\alpha^2)^{1/2}, \quad \lambda \equiv (1 - E^2)^{1/2}, \quad \Delta \equiv (E_0 - E), \quad E_0 \equiv (1 - Z^2\alpha^2)^{1/2}, \end{aligned} \tag{2.14}$$

where E is the energy of the GSHA perturbed by a finite nuclear size and E_0 is the unperturbed energy of the GSHA.

Establishing the legitimacy of the singular solution of the Dirac equation outside the nucleus (for potentials satisfying the matching condition (2.13)) was *the fundamental discovery in its own right*—with potential consequences far beyond the puzzle of the HTMD in the GSHA. Being applied to the latter problem, this fundamental discovery bridged the gap between the experimental HTMD and the previous theoretical HTMD. Indeed, as we explained by analyzing equation (2.4), a strong singularity of the radial part of the coordinate wave function $R_{10}^{\text{true}}(r)$ at small r leads to a much slower decline of the HTMD at large momenta p. The huge discrepancy is eliminated.

This fundamental result was derived from standard quantum mechanics. In [3] the author did not change any physical laws.

In [1] the author significantly advanced the results from [3], as presented below. For a system consisting of an electron in the Coulomb field there are four operators that have common eigenfunctions. These operators are the square of the total angular momentum J^2, the projection J_z of the total angular momentum, the Hamiltonian H, as well as the following operator:

$$K = \beta(2\mathbf{Ls} + 1). \tag{2.15}$$

In equation (2.15), **L** and **s** are the operators of the orbital angular momentum and spin, respectively, **Ls** is their dot-product (the scalar product), β is a Dirac matrix of rank four. The nonzero elements of the matrix β are $\beta_{11} = \beta_{22} = 1$, $\beta_{33} = \beta_{44} = -1$. The eigenvalues of the operator J^2 and the eigenvalue of the operator K are interrelated: $k = \pm(j + 1/2)$.

For hydrogen atoms, the eigenvalues of the Hamiltonian (i.e. the energy) are

$$E_{Nk} = mc^2\left\{1 + \alpha^2/[N + (k^2 - \alpha^2)^{1/2}]^2\right\}^{-1/2}, \tag{2.16}$$

where N is the radial quantum number. The ground state is characterized by

$$N = 0,\ k = -1. \tag{2.17}$$

Therefore

$$E_{0,-1} = mc^2(1 - \alpha^2)^{1/2}. \tag{2.18}$$

The radial part $R_{Nk}(r)$ of the coordinate wave function at small r is (see e.g. [11])

$$R_{Nk}(r) \propto 1/r^{1+s}, \qquad s = \pm(k^2 - \alpha^2)^{1/2}. \tag{2.19}$$

For the ground state we obtain

$$R_{0,-1}(r) \propto 1/r^q, \qquad q = 1 \pm (1 - \alpha^2)^{1/2}. \tag{2.20}$$

The central point is that the main result of [3]—the statement that the singular exterior solution (i.e. outside the proton) corresponding to $q = 1 + (1 - \alpha^2)^{1/2}$ is appropriate for the ground state—was derived employing *only* the fact that in the ground state the eigenvalue of the operator K is $k = -1$. However, there is an infinite number of other states characterized by $k = -1$. These states have zero value of the orbital angular momentum ($l = 0$)—they are S-states, more specifically $^2S_{1/2}$ states. The consequence of this result was formulated by the author in [1] as follows: 'Both the regular exterior solution corresponding to $q = 1 - (1 - \alpha^2)^{1/2}$ and the singular exterior solution corresponding to $q = 1 + (1 - \alpha^2)^{1/2}$ are legitimate not only for the ground state $1^2S_{1/2}$, but also for the states $2^2S_{1/2}$, $3^2S_{1/2}$, and so on, i.e. for the states $n^2S_{1/2}$, where $n = N + |k| = N + 1$ is the principal quantum number ($n = 1, 2, 3, \ldots$). Both the regular exterior solution corresponding to $q = 1 - (1 - \alpha^2)^{1/2}$ and the singular exterior solution corresponding to $q = 1 + (1 - \alpha^2)^{1/2}$ are legitimate also for the $l = 0$ states of the continuous spectrum. Thus, according to these theoretical results, there is an alternative kind of hydrogen atom (AKHA), corresponding to the singular solution outside the proton.'

Because the AKHA have only S-states, then according to the selection rules they practically do not interact with the electromagnetic radiation. They remain dark (or almost dark). This fundamental discovery turned out to have far reaching implications, including consequences for understanding the composition of the Universe, as explained in the next section.

2.2 An alternative kind of hydrogen atom as a possible explanation of the latest puzzling observation of the 21 cm radio line from the early Universe

Studying signals from the Universe in the radio frequency range plays a very important role in understanding the Universe. One of the most informative parts of radio astronomy relates to the 21 cm radio line of hydrogen.

The 21 cm radio line corresponds to the radiative transition between very closely spaced hyperfine structure sublevels of the ground state of hydrogen atoms. These sublevels differ very slightly in energy depending on the mutual orientation of the electron spin and the proton spin. Without any external influence on a hydrogen atom, it would take about 10 million years for the transition between the hyperfine structure sublevels to occur. However, in the early Universe, as the ultraviolet backlighting from newborn stars penetrated huge clouds of neutral hydrogen, it caused the excitation of the upper hyperfine structure sublevel and the corresponding radiative transition. For observers on Earth, this electromagnetic transition is very significantly redshifted compared to its original wavelength of 21 cm.

In 2018 Bowman *et al* [23], while observing the 21 cm line from the early Universe, encountered a puzzling result: the amplitude of the absorption profile of this (redshifted) line was more than a factor of two greater than expected from the standard cosmology. This could have been translated into the statement that the gas temperature in the hydrogen clouds was significantly smaller than predicted by the standard cosmology [23].

The first possible explanation was proposed by Barkana [24]. He assumed that collisions with some kind of unspecified dark matter could have been making the hydrogen gas cooler than expected. He estimated that the particles of this dark matter should be lighter than 4.3 GeV.

Others proposed hypothetical alternative explanations, such as those in [25–27]. However, their hypotheses were also quite exotic.

In [1] the author was motivated by the following question: what if instead of Barkana's unspecified dark matter, it was the AKHA, whose existence was already proven by the analysis of atomic experiments (as described above)?

The AKHA do not possess excited states that can be coupled to the ground state by electric-dipole radiation, this being a striking difference to ordinary hydrogen atoms. (The AKHA do have two hyperfine sublevels of the ground state, as the usual species of hydrogen atoms.) In [1] the author showed that because of this, the AKHA decouple from the cosmic microwave background radiation earlier than ordinary hydrogen atoms. This lowers the temperature of the hydrogen gas sufficiently to explain the puzzling observation by Bowman *et al* [23] not only qualitatively, but also quantitatively, as shown in detail in [2].

The author wrote [1]:

> This explanation seems to be more specific and natural than adopting a possible cooling of baryons either by unspecified dark matter particles, as in [the] paper by Barkana [23], or by some exotic dark matter particles [with a] charge [a] million times smaller than the electron charge, as in [the] paper by Muñoz and Loeb [27]. Also our explanation does not require an additional radio background [as] suggested by Feng and Holder [25] and by Ewall-Wice *et al* [26]. Further observational studies of the redshifted 21 cm radio line from the early Universe could help to find out which explanation is the most relevant.

The above results constitute a possible answer to one of the most captivating questions in cosmology: dark matter (or a part of it) is the AKHA. The fundamental results on the existence and the structure of the AKHA were derived in papers [1, 3] from standard quantum mechanics. Thus, while explaining the puzzling astrophysical observations by Bowman *et al* [23], the author did not change any physical laws [1]—in distinction to numerous unproven hypotheses about dark matter that resort to changing physical laws. Also, the explanation in [1] did not require assuming never-discovered subatomic particles—in distinction to numerous papers dealing with those hypothetical particles, despite the absence of any definitive experimental evidence for these subatomic particles—see the sketch in figure 2.2.

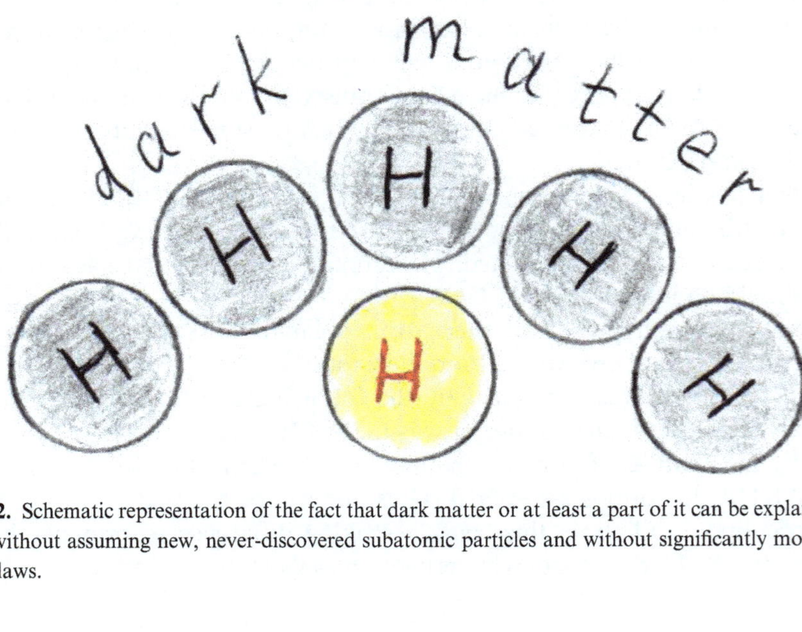

Figure 2.2. Schematic representation of the fact that dark matter or at least a part of it can be explained by the AKHA without assuming new, never-discovered subatomic particles and without significantly modifying the physical laws.

In distinction to all other hypotheses about dark matter, the existence of the AKHA has already been proven by the analysis of atomic experiments (as explained above). Therefore, the concept of the AKHA as dark matter constitutes a breakthrough in understanding the 'nuts and bolts' of the Universe.

2.3 Two flavors of hydrogen atoms

In [2] the author focused on the logical consequences of the fundamental discovery from [3]. Let us look at the results from [3] from the following point of view.

The ground state of the usual hydrogen atoms and the ground state of the AKHA have the same energy and correspond to the same pair of quantum numbers: $N = 0$, $k = -1$. Clearly their ground state has an *additional double-degeneracy*.

The cause of any degeneracy in quantum mechanics is well-understood and it has to do with commutators of quantum operators. If there are operators A_i, such that the commutator $\{A_i, H\} = 0$ (so that the physical quantities corresponding to A_i are conserved), but the commutator $\{A_i, A_j\}$ is nonzero, then the eigenvalues of the Hamiltonian (i.e. the energy), generally speaking, are degenerate. This fundamental theorem works both ways: if there is a degeneracy, then the system should have a conserved quantity (or quantities) whose operator (or operators) do not commute with the Hamiltonian. Any additional degeneracy entails the existence of yet another conserved quantity, whose operator does not commute with the Hamiltonian.

Therefore, the fact of the additional double degeneracy of the ground state of hydrogen atoms means that for hydrogen atoms *there exists an additional, new conserved quantity*. The ground state of the usual hydrogen atoms and the ground state of the AKHA differ by eigenvalues of this previously unknown conserved quantity (while eigenvalues of the known conserved quantities E, K, J^2, and J_z are the same for both ground states). From the results of [2, 3] it follows that this additional double-degeneracy relates not only to the ground states of hydrogen atoms, but to all their S-states, including the excited states.

Let us now make an excursion to the realm of quarks. It is well-known that quarks have been assigned *flavors*. This was actually a shortcut for stating that there are pairs of quarks that differ by the eigenvalue of some physical quantity. For example, to the pair up-quark and down-quark was assigned an operator of the isotopic spin (isospin) I. The eigenvalue $I_z = 1/2$ of its z-projection was assigned to the up-quark and the eigenvalue $I_z = -1/2$ of its z-projection was assigned to the down-quark. It is said that quarks have *flavor symmetry*.

The above situation with two kinds of hydrogen atoms is analogous. Therefore, it is appropriate to state that hydrogen atoms have *flavor symmetry*, similarly to quarks. Hydrogen atoms have *two flavors* that can be distinguished by the eigenvalue of an additional, new conserved quantity, to which corresponds a new operator.

In [2] the author called this new conserved quantity *isohydrogen spin* and denoted it as $I^{(h)}$. (He also suggested an abbreviated name: *isohyspin*.) In analogy to the isospin of up- and down-quarks, the eigenvalue $I_z^{(h)} = 1/2$ of its z-projection is assigned to the first flavor of hydrogen atoms (i.e. to the ordinary hydrogen atoms) and the eigenvalue $I_z^{(h)} = -1/2$ is assigned to the second flavor of hydrogen atoms (i.e. to the AKHA).

Finally, the author emphasized the following [2]. The concept of the isohyspin, while being important from the fundamental point of view, is not needed to confirm the AKHA as dark matter or a part of it.

References

[1] Oks E 2020 *Res. Astron. Astrophys.* **20** 109

[2] Oks E 2020 *Atoms* **8** 33

[3] Oks E 2001 *J. Phys.* B **34** 2235

[4] Oks E 2015 *Breaking Paradigms in Atomic and Molecular Physics* (Singapore: World Scientific)

[5] Fock V 1935 *Z. Phys.* **98** 145

[6] Gryzinski M P 1963 *6th Int. Conf. On Ionization Phenomena in Gases (Paris)* p 117

[7] Gryzinski M 1965 *Phys. Rev.* **138** A336

[8] Flügge S 1974 *Practical Quantum Mechanics* (New York: Springer) ch II.D

[9] Landau L D and Lifshitz E M 1965 *Quantum Mechanics* (Oxford: Pergamon)

[10] Bialynicki-Birula I, Cieplak M and Kaminski J 1992 *Theory of Quanta* (New York: Oxford University Press)

[11] Rose M E 1961 *Relativistic Electron Theory* (New York: Wiley) sects 28, 29, 39

[12] Greiner W 1994 *Relativistic Quantum Mechanics, Wave Equations* (New York: Springer) sect 9

[13] Flügge S 1974 *Practical Quantum Mechanics* (New York: Springer) problem 202

[14] Berestetskii V B, Lifshitz E M and Pitaevskii P 1971 *Relativistic Quantum Theory* (Oxford: Pergamon) sect 36

[15] Pomeranchuk I and Smorodinskii J 1945 *J. Phys. USSR* **9** 97

[16] Werner F G and Wheeler J A 1958 *Phys. Rev.* **109** 126

[17] Rein D 1969 *Z. Phys.* **221** 423

[18] Popov V S 1971 *Sov. J. Nucl. Phys.* **12** 235

[19] Greiner W and Reinhardt J 1994 *Quantum Electrodynamics* (New York: Springer) sect 7.1

[20] Simon G G, Schmitt C, Borkowski F and Walther V H 1980 *Nucl. Phys.* A **333** 381

[21] Perkins D H 1987 *Introduction to High Energy Physics* (Menlo Park, CA: Addison-Wesley) sect 6.5

[22] Feynman R P 1989 *Photon–Hadron Interactions* (Redwood City, CA: Addison-Wesley) lecture 24

[23] Bowman J D, Rogers A E E, Monsalve R A, Mozdzen T J and Mahesh N 2018 *Nature* **555** 67

[24] Barkana R 2018 *Nature* **555** 71

[25] Feng C and Holder J 2018 *Astrophys. J. Lett.* **858** L17

[26] Ewall-Wice A, Chang T-C, Lazio J, Doré O, Seiffert M and Monsalve R A 2018 *Astrophys. J.* **868** 63

[27] Muñoz J B and Loeb A 2018 *Nature* **557** 684

IOP Publishing

Simple Atomic and Molecular Systems
New results and applications
Eugene Oks

Chapter 3

Monopole contribution to the Stark width of hydrogen-like spectral lines in plasmas

A sample of the background literature on the subject can be found in [1–8] (and references therein). Here we follow [9].

According to equation (3b) from [8], the monopole contribution to the interaction potential has the form

$$V^{(0)}(t) = -e^2[1/|R(t) - 1/r]\, E[R(t) < r]. \tag{3.1}$$

Here $E[\ldots]$ is the Heaviside function and R and r are the absolute values of the radius-vectors of the perturbing electron and of the bound electron, respectively. For the Lyman lines, in a plasma with electron density N_e and electron velocity v, the shift due to the monopole interaction is (see equation (17) from [8]):

$$d_{nl\to 1s} = 2\pi N_e v \int_0^{\rho_{max}} \rho \sin[\langle nl|\Phi_0|nl\rangle - \langle 1s|\Phi_0|1s\rangle]\, d\rho. \tag{3.2}$$

The matrix elements of the electron broadening operator entering equation (3.2) are as follows:

$$\begin{aligned}&\langle nlm|\Phi_0|nl'm'\rangle = \\ &-[e^2/(\hbar v)](1 + u_0)\{\ln[(1+x)/(1-x)] - 2x\}\, E[R(t) < r_{nl}]\, \delta_{ll'}\delta_{mm'}.\end{aligned} \tag{3.3}$$

Here

$$x = [1 - (u^2 + u_0{}^2)/(1+u_0)^2]^{1/2}, \quad u = \rho/r_{nl}, \quad u_0 = \rho_0/r_{nl}, \tag{3.4}$$

where ρ is the impact parameter and

$$\begin{aligned}\rho_0 = (Z-1)e^2/(m_e v^2), \qquad r_{nl} &= (\langle nl|r^2|nl\rangle)^{1/2} \\ &= (a_0 n/Z)\{[5n^2 + 1 - 3l(l+1)]/2\}^{1/2}.\end{aligned} \tag{3.5}$$

Here a_0 is the Bohr radius and r_{nl} is the root-mean-square size of the radiator in the state characterized by the quantum numbers n and l. Some practical formulas from [8] can be useful:

$$e^2/(\hbar v) = [13.605/kT_e\,(\text{eV})]^{1/2}, \quad u_0 = [Z(Z-1)/n^2]\,[13.605/kT_e\,(\text{eV})]. \quad (3.6)$$

Figure 3.1 shows the dependence of x from equation (3.4) on u and u_0.

From the condition $R < r_{nl}$, one obtains

$$u_{\max} = \rho_{\max}/r_{nl} = (1 + 2u_0)^{1/2}. \quad (3.7)$$

Figure 3.2 illustrates the dependence of $u_{\max}$ on u_0.

Equation (3.7) is equivalent to

$$\rho_{\max} = (r_{nl}{}^2 + 2\,r_{nl}\rho_0)^{1/2}. \quad (3.8)$$

Figure 3.3 shows the dependence of $\rho_{\max}$ on r_{nl} and ρ_0.

For relatively high temperatures, such that $e^2/(\hbar v) \ll 1$, one has $|\langle nlm|\Phi_0|nlm\rangle| < 1$, as noted in [8]. In the opposite case, where the temperature is relatively low ($u_0 \gg 1$), the matrix element $|\langle nlm|\Phi_0|nlm\rangle|$ does not exceed $(4/3)2^{1/2}n/[Z(Z-1)]^{1/2}$, as estimated in [8]. The authors of [8] replaced sin[…] by its argument in equation (3.2). This was justified for the situation where n is no more than 4 and Z is no less than 5.

In this chapter we present the monopole contribution to the width $w^{(0)}$. For the Lyman lines we use equation (3.2) where we replace sin[…] by cos[…]:

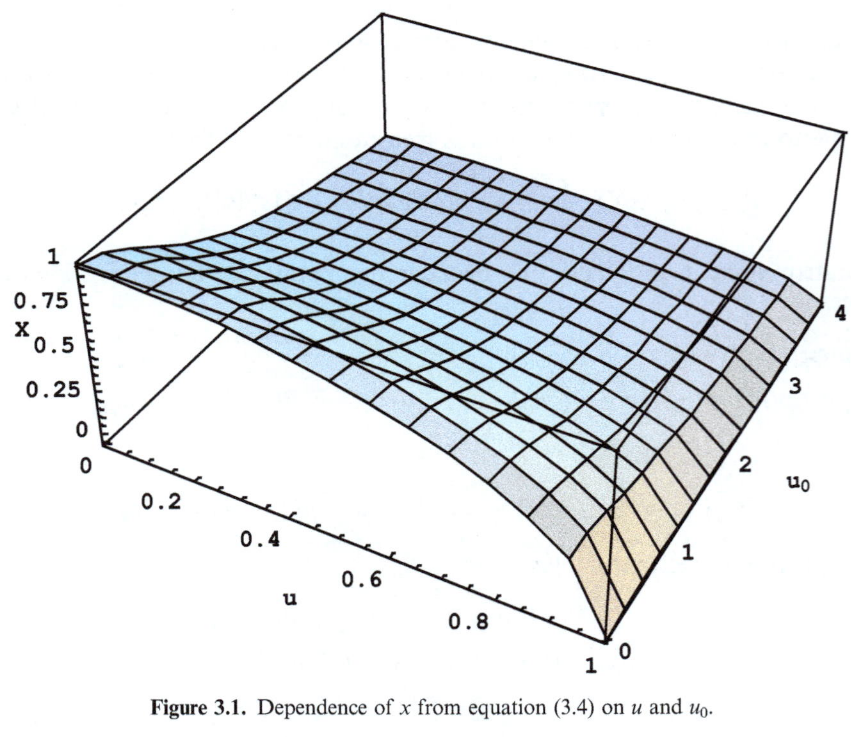

Figure 3.1. Dependence of x from equation (3.4) on u and u_0.

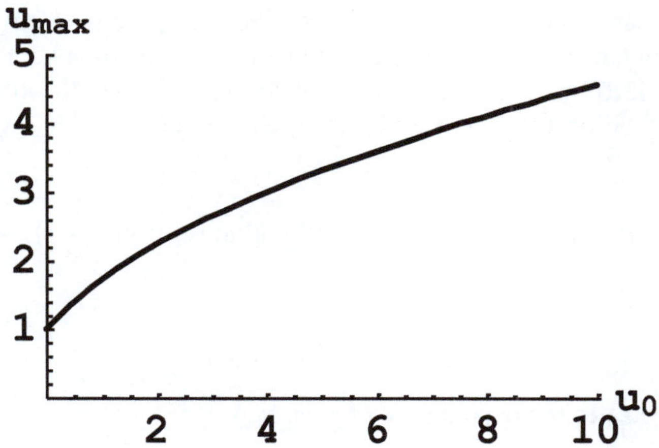

Figure 3.2. Dependence of $u_{\max}$ from equation (3.7) on u_0.

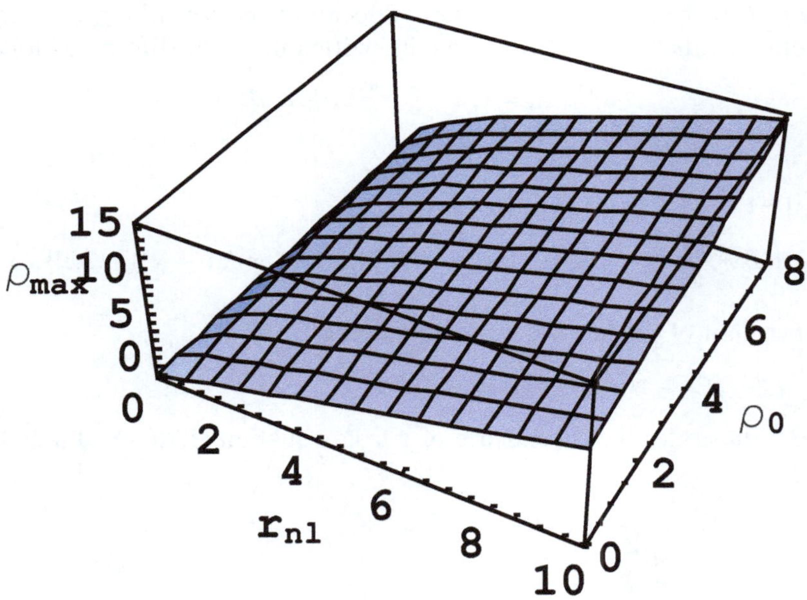

Figure 3.3. Dependence of $\rho_{\max}$ from equation (3.8) on r_{nl} and ρ_0.

$$w^{(0)}{}_{nl\to 1s} = 2\pi N_e v \int_0^{\rho_{\max}} \rho \cos\left[\langle nl|\Phi_0|nl\rangle - \langle 1s|\Phi_0|1s\rangle\right] \, d\rho. \tag{3.9}$$

We focus on the situation where the contribution of the ground level can be disregarded. This is legitimate for $n \gg 1$ (or practically $n > 4$). Then equation (3.9) simplifies to

$$w^{(0)}{}_{nl\to 1s} = 2\pi N_e v \int_0^{\rho_{\max}} \rho \cos\left(\langle nl|\Phi_0|nl\rangle\right) d\rho. \tag{3.10}$$

We do not assume that $|\langle nlm|\Phi_0|nlm\rangle| < 1$. Therefore, the corresponding trigonometric function (cos [...]) in the integrand in equation (3.10) is kept here.

In equation (3.10) we proceed from the integration over ρ to the integration over x (by using the relation between x and ρ from equation (3.4)):

$$w^{(0)}{}_{nl\to 1s} = 2\pi N_e v r_{nl}{}^2(1 + u_0)^2 \int_0^y x \cos\{[e^2(1 + u_0)/(\hbar v)][\ln((1 + x)/(1 - x)) - 2x]\}, \tag{3.11}$$

where

$$y = (1 + 2u_0)^{1/2}/(1 + u_0). \tag{3.12}$$

Figure 3.4 illustrates the dependence of y on u_0.

We denote

$$A = (Z - 1)a_B/r_{nl}, \quad B = \hbar v/e^2. \tag{3.13}$$

Physically B is the scaled dimensionless velocity of the perturbing electrons. With these notations the width $w^{(0)}{}_{nl\to 1\ s}$ can be written in the final form, as follows:

$$w^{(0)}{}_{nl\to 1s} = (2\pi N_e r_{nl}{}^2 e^2/\hbar)F[A, B]. \tag{3.14}$$

Here

$$F[A, B] = B(1 + A/B^2)^2 \int_0^{y(A,\,B)} x \cos\{(1/B + A/B^3)\,[\ln((1 + x)/(1 - x)) - 2x]\}\,\mathrm{d}x. \tag{3.15}$$

The upper limit of the integration in equation (3.15) is

$$y(A, B) = (1 + 2A/B^2)^{1/2}/(1 + A/B^2). \tag{3.16}$$

Figure 3.5 illustrates the dependence of y from equation (3.16) on A and B.

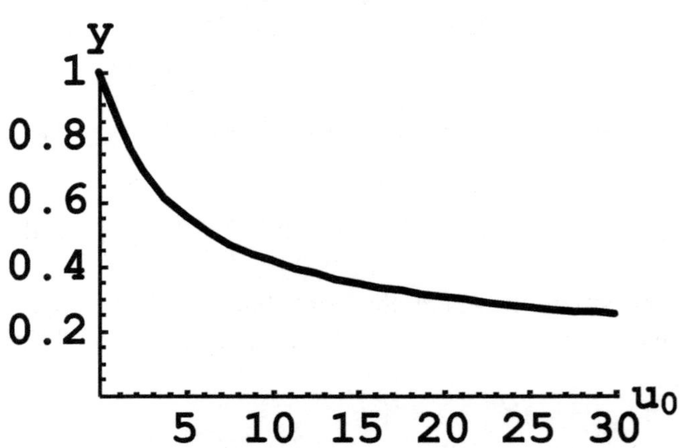

Figure 3.4. Dependence of y from equation (3.12) on u_0.

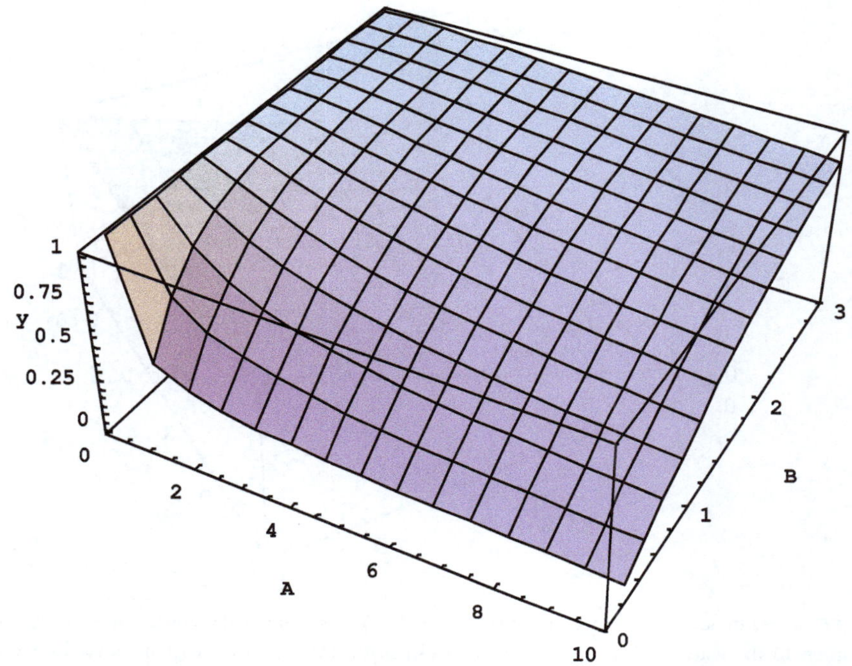

Figure 3.5. Dependence of y from equation (3.16) on parameters A and B defined in equation (3.13).

Thus, the function $F(A, B)$ provides the dependence of the width $w^{(0)}{}_{nl\to 1\ s}$ on the scaled dimensionless electron velocity B. Figure 3.6 presents a three-dimensional plot of this function.

Figure 3.7 illustrates the dependence of the function $F(A, B)$ on the scaled dimensionless electron velocity B for three values of the parameter A: $A = 1$ (solid line), $A = 0.6$ (dashed line), and $A = 0.3$ (dash-dotted line). Figures 3.6 and 3.7 clearly show that the width $w^{(0)}{}_{nl\to 1\ s}$ has a non-monotonic dependence on B. It is a *counter-intuitive result* that the monopole contribution to the widths has the non-monotonic dependence on the electron velocity.

When $B \gg \max(A, 1)$, i.e. when electron velocities are relatively large, it becomes easy to perform the integration in equation (3.15). We obtain $F(A, B) = B/2$, so that

$$w^{(0)}{}_{nl\to 1s} = \pi r_{nl}{}^2 N_e v. \tag{3.17}$$

The physical meaning is that the cross-section for the line broadening collisions in this case becomes equal to the 'geometrical' cross-section $\pi r_{nl}{}^2$.

When electron velocities are relatively large, then as the velocity increases so does the monopole contribution to the width. This is *very different* from the dipole contribution to the width. The latter diminishes as the electron velocity increases.

The parameter B (defined in equation (3.13)) does not reach the range of $B \gg 1$ for thermal velocities of plasma electrons. However, a relativistic electron beam (REB) is encountered in some plasma experiments, such as magnetic fusion, inertial fusion, acceleration of charged particles in plasmas, and generation of high-intensity coherent microwave radiation—we refer to e.g. [10–12] and references therein.

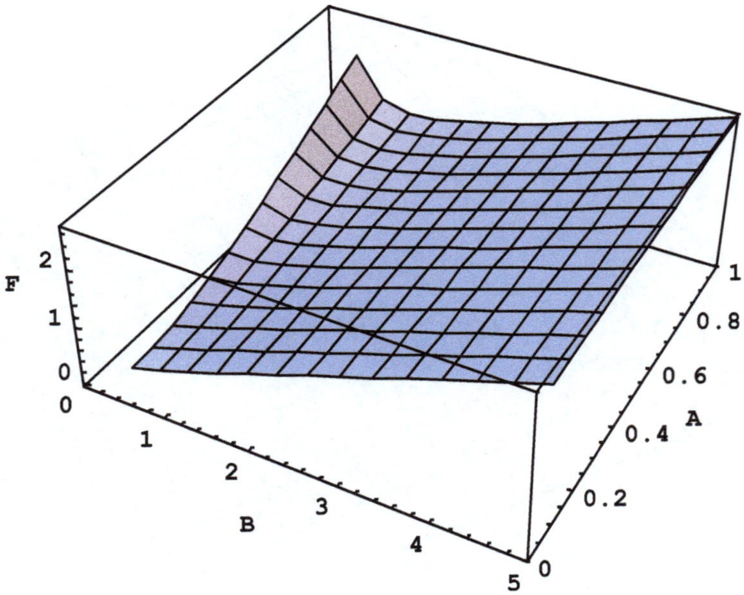

Figure 3.6. Three-dimensional plot of the function $F(A, B)$ representing the dependence of the monopole contribution to the width $w^{(0)}{}_{nl\to 1s}$ (from equation (3.14)) on the scaled dimensionless electron velocity B (defined in equation (3.13)). Reproduced from [9] with permission of MDPI. The function $F(A, B)$ is defined by equations (3.15) and (3.16).

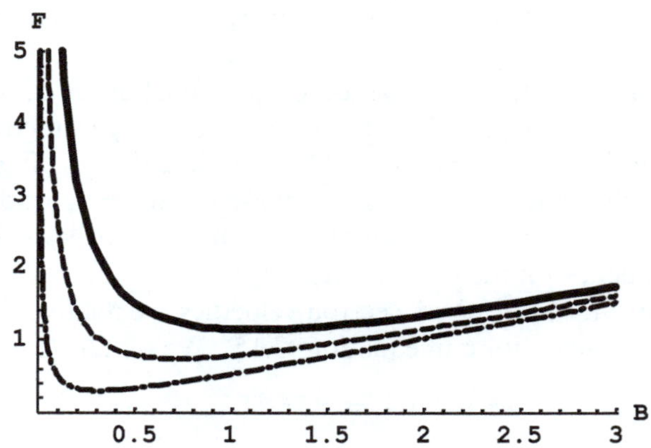

Figure 3.7. Plot of the function $F(A, B)$ representing the dependence of the monopole contribution to the width $w^{(0)}{}_{nl\to 1s}$ (from equation (3.14) on the scaled dimensionless electron velocity B (defined in equation (3.13))) for three values of the parameter A: $A = 1$ (solid line), $A = 0.6$ (dashed line), and $A = 0.3$ (dash-dotted line). Reproduced from [9] with permission of MDPI. The function $F(A, B)$ is defined by equations (3.15) and (3.16).

Specifically, so-called runaway electrons are encountered in magnetic fusion research experiments. In tokamaks, the plasma current—in some discharges—decays and is partly replaced by runaway electrons. The runaway electrons can reach relativistic energies. This represents a serious risk to the function of the next generation tokamak ITER—see e.g. [13–15] and references therein.

In one of our papers [16], the *dipole* contribution w_d to the Stark width of hydrogenic spectral lines due to an REB has been calculated analytically and estimated as follows:

$$w_d \sim N_{beam}c(n^2/Z)^2\lambda_{Comp}{}^2/(1 - 1/\gamma^2)^{1/2}, \quad \lambda_{Comp} = \hbar/(m_e c), \quad \gamma = 1/(1 - v^2/c^2)^{1/2}. \tag{3.18}$$

In equation (3.18) λ_{Comp} is the Compton wavelength, N_{beam} is the REB density, and γ is the relativistic factor. By employing equations (3.5), (3.17), and (3.18) it is possible to estimate the ratio of the corresponding *monopole* contribution w_m due to the REB to w_d as follows:

$$w_m/w_d \sim (a_0/\lambda_{Comp})^2/(1 - 1/\gamma^2)^{1/2}. \tag{3.19}$$

For $\gamma \gg 1$, i.e. for an ultra-relativistic REB, equation (3.19) reduces to

$$w_m/w_d \sim (a_0/\lambda_{Comp})^2 = (\hbar c/e^2)^2 \sim 10^4 \gg 1. \tag{3.20}$$

Thus the monopole contribution to the width due to the REB in this situation is greater than the corresponding dipole contribution by four orders of magnitude and practically determines the entire Stark width of hydrogenic spectral lines due to the REB.

References

[1] Griem H R and Shen K Y 1961 *Phys. Rev.* **122** 1490
[2] Griem H R 1964 *Plasma Spectroscopy* (New York: McGraw-Hill)
[3] Griem H R 1974 *Spectral Line Broadening by Plasmas* (Cambridge, MA: Academic)
[4] Sanders P and Oks E 2018 *J. Phys. Commun.* **2** 035033
[5] Oks E, Derevianko A and Ispolatov Y 1995 *J. Quant. Spectrosc. Rad. Transfer* **54** 307
[6] Oks E 2006 *Stark Broadening of Hydrogen and Hydrogenlike Spectral Lines in Plasmas: The Physical Insight* (Oxford: Alpha Science International)
[7] Oks E 2017 *Diagnostics of Laboratory and Astrophysical Plasmas Using Spectral Lines of One- Two-, and Three-Electron Systems* (Hackensack, NJ: World Scientific)
[8] Mejri M A, Nguyen H and Ben Lakhdar Z 1998 *Eur. Phys. J.* D **4** 125
[9] Oks E 2020 *Plasma* **3** 180
[10] Guenot D *et al* 2017 *Nat. Photon.* **11** 293
[11] Kurkin S A, Hramov A E and Koronovskii A A 2013 *Appl. Phys. Lett.* **103** 43507
[12] de Jagher P C, Sluijter F W and Hopman H J 1988 *Phys. Rep.* **167** 177
[13] Boozer A H 2017 *Nucl. Fusion* **57** 056018
[14] Decker J, Hirvijoki E, Embreus O, Peysson Y, Stahl A, Pusztai I and Fülöp T 2016 *Plasma Phys. Control. Fusion* **58** 025016
[15] Smith H, Helander P, Eriksson L-G, Anderson D, Lisak M and Andersson F 2006 *Phys. Plasmas* **13** 102502
[16] Oks E and Sanders P 2018 *J. Phys. Commun.* **2** 015030

Chapter 4

How the finite mass of nuclei complicates analytical treatments of hydrogen atoms in external fields

4.1 Hydrogen atoms in a magnetic field

There are lots of works demonstrating that for hydrogenic atoms/ions in a uniform magnetic field, the center-of-mass motion and the relative (internal) motion are coupled by the magnetic field. Therefore, these two kinds of motion, generally speaking, cannot be separated—see e.g. [1–3] and references therein. A pseudoseparation is possible for hydrogen atoms (but not for hydrogen-like ions). It leads to a Hamiltonian for the relative motion that depends on a center-of-mass vector integral of the motion **K** called pseudomomentum [3].

For hydrogen atoms, the components of the pseudomomentum vector commute with each other. This allows the pseudoseparation of the center-of-mass motion and the relative motion. The influence of the center-of-mass motion on the relative motion is reduced physically to the motional Stark effect.

The diamagnetic potential term in the Hamiltonian for the relative motion leads to the formation of an additional potential well. This potential well is located far away from the hydrogen nucleus (proton). When the magnetic field is relatively strong, the new bound states inside this well are delocalized states having almost macroscopic dimensions. Thus, the bound state inside this well is characterized by a very large electric dipole moment. Therefore, such states are very sensitive to an external electric field.

In the Hamiltonian of the relative (internal) motion for the hydrogen atom in a magnetic field **B**, the potential energy has the form (see e.g. [4] equation (4.6))

$$V = [e^2/(2Mc)]\mathbf{B} \times \mathbf{r} - [|e|/(Mc](\mathbf{B} \times \mathbf{K})\mathbf{r} - e^2/r, \qquad (4.1)$$

doi:10.1088/978-0-7503-3679-6ch4

where **K** is the pseudomomentum, M is the mass of the hydrogen atom, c is the speed of light, and e is the electron charge. Just as in [4], we choose $e < 0$. It should also be mentioned that in [4] the authors set $c = 1$. In equation (4.1) $(\mathbf{B} \times \mathbf{K})\mathbf{r}$ stands for the scalar product (also known as the dot-product) of vector $\mathbf{r}$ and vector $(\mathbf{B} \times \mathbf{K})$.

We consider the same configuration as presented in [4]: $\mathbf{B} = (0, 0, B)$ and $\mathbf{K} = (0, K, 0)$. The location of the additional potential well is determined by equating $\mathrm{d}V/\mathrm{d}\mathbf{r} = 0$, from where we obtain

$$y = z = 0, \tag{4.2}$$

$$x^3 + [cK/(|e|B]x^2 + Mc^2/B^2 = 0. \tag{4.3}$$

After introducing the notations

$$b = B/(cM^{1/2}), \qquad k = K/(|e|M^{1/2}), \tag{4.4}$$

we simplify equation (4.3) as follows:

$$x^3 + (k/b)x^2 + 1/b^2 = 0. \tag{4.5}$$

Out of the three roots of the cubic equation (4.5) two roots are always complex. Under the condition

$$k > 3(b/4)^{1/3} \tag{4.6}$$

the third root, denoted here as x_0, is real and is located at the negative part of the x-axis. Figure 4.1 shows the dependence of the location x_0 of the additional potential well on the scaled magnetic field b.

Schmelcher and Cederbaum [4] proposed only an approximate formula for x_0. In our notations, their approximate formula reads:

$$x_{0,\mathrm{CS}} = -k/b + k/(k^3 - 2b). \tag{4.7}$$

As an example, figure 4.2 presents the ratio of the exact root x_0 from equation (4.5) to the approximate root $x_{0,\mathrm{CS}}$ from [4] for the scaled pseudomomentum $k = 15$.

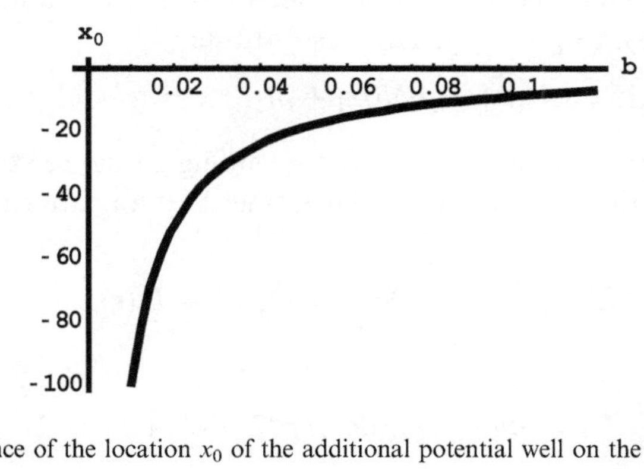

Figure 4.1. Dependence of the location x_0 of the additional potential well on the scaled magnetic field b.

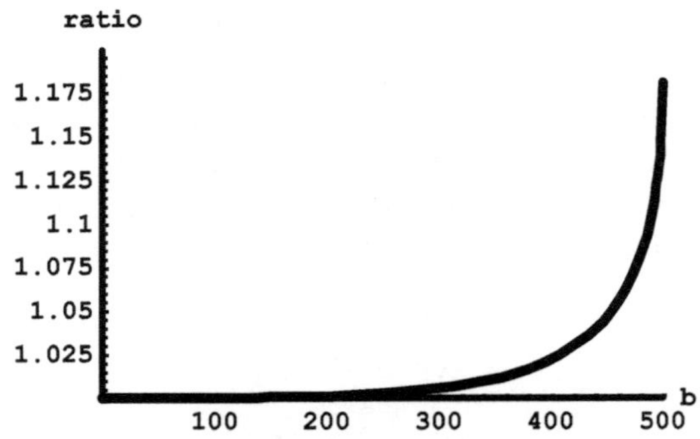

Figure 4.2. The ratio of the exact root x_0 from equation (4.5) to the approximate root $x_{0,\text{CS}}$ from [4] reproduced in equation (4.7) for the scaled pseudomomentum $k = 15$.

It is seen that for a given k, the relative error of the approximate formula (4.7) from [4] grows larger as the scaled magnetic field b increases.

For hydrogen-like ions in a magnetic field the situation is more complicated than for hydrogen atoms. In the case of hydrogen-like ions, the pseudomomentum vector is conserved but its components do not commute. Therefore, for hydrogen-like ions the components of the pseudomomentum cannot be employed for the pseudoseparation of the center-of-mass motion and the relative motion. Further details can be found in e.g. [5].

4.2 Hydrogen atoms in an electric field

For hydrogen atoms and hydrogen-like ions in a *uniform* electric field, the relative motion and the center-of-mass motion can be exactly separated—see e.g. [6]. In [7, 8] the author analyzed this situation with respect to a *non-uniform* electric field. Here, we present the results from [7, 8].

We consider a system of two charges e_1 and e_2 of masses m_1 and m_2, respectively, situated in a non-uniform electric field. The Lagrangian of the system can be represented in the form

$$L = [m_1(d\mathbf{r}_1/dt)^2 + m_2(d\mathbf{r}_2/dt)^2]/2 - e_1e_2/|\mathbf{r}_2 - \mathbf{r}_1| - e_1\varphi(\mathbf{r}_1) - e_2\varphi(\mathbf{r}_2). \tag{4.8}$$

Here φ is the potential of the non-uniform electric field and $\mathbf{r}_1$ and $\mathbf{r}_2$ are radii vectors of charges e_1 and e_2, respectively. Let us substitute

$$\mathbf{R} = (m_1\mathbf{r}_1 + m_2\mathbf{r}_2)/(m_1 + m_2), \quad \mathbf{r} = \mathbf{r}_2 - \mathbf{r}_1, \tag{4.9}$$

where $\mathbf{r}$ and $\mathbf{R}$ are the radius-vectors corresponding to the relative motion and the center-of-mass motion, respectively. Then the Lagrangian can be rewritten as follows:

$$L(\mathbf{R}, \mathbf{r}) = L_{\text{CM}}(\mathbf{R}) - U(\mathbf{R}, \mathbf{r}) + L_r(\mathbf{r}). \tag{4.10}$$

Here

$$L_{\text{CM}}(\mathbf{R}) = (m_1 + m_2)(d\mathbf{R}/dt)^2/2 - (e_1 + e_2)\varphi(\mathbf{R}) \tag{4.11}$$

is the Lagrangian of the center-of-mass,

$$L_r(\mathbf{r}) = \mu(\mathrm{d}\mathbf{r}/\mathrm{d}t)^2/2 - e_1e_2/r \tag{4.12}$$

is the Lagrangian of the relative motion, and

$$U(\mathbf{R}, \mathbf{r}) = \mu(e_1/m_1 - e_2/m_2)\mathbf{r}\mathbf{F}(\mathbf{R}) \tag{4.13}$$

is the coupling of the center-of-mass motion and relative motion. In equations (4.12) and (4.13)

$$\mu = m_1m_2/(m_1 + m_2) \tag{4.14}$$

is the reduced mass of the two particles and

$$\mathbf{F}(\mathbf{R}) = -\mathrm{d}\varphi(\mathbf{R})/\mathrm{d}\mathbf{R} \tag{4.15}$$

is a non-uniform electric field.

The following Hamiltonian corresponds to the Lagrangian from equation (4.10):

$$H = H_{\mathrm{CM}}(\mathbf{R}, \mathbf{P}) + U(\mathbf{R}, \mathbf{r}) + H_r(\mathbf{r}, \mathbf{p}), \tag{4.16}$$

where

$$H_{\mathrm{CM}}(\mathbf{R}, \mathbf{P}) = P^2/[2(m_1 + m_2)] + (e_1 + e_2)\varphi(\mathbf{R}) \tag{4.17}$$

is the Hamiltonian of the center-of-mass and

$$H_r(\mathbf{r}, \mathbf{p}) = p^2/(2\mu) + e_1e_2/r \tag{4.18}$$

is the Hamiltonian of the relative motion. In equations (4.17) and (4.18) $\mathbf{p}$ is the momentum of the relative motion and $\mathbf{P}$ is the momentum of the center-of-mass motion.

Equations (4.11)–(4.13) and (4.16)–(4.18) demonstrate that the relative motion and center-of-mass motion and they are coupled in a non-uniform electric field—specifically by the term $U(\mathbf{R}, \mathbf{r})$ from equation (4.13). Thus the relative motion and the center-of-mass motion in a non-uniform electric field cannot be separated in the general case.

Now, let us consider the situation where $m_1 \ll m_2$. In this situation, by using the approximate analytical method of separating rapid and slow subsystems, the relative motion and the center-of-mass motion and can be separated. The characteristic frequency of the relative motion in this case is much greater than the characteristic frequency of the center-of-mass motion. The relative motion corresponds to the rapid subsystem, while the center-of-mass motion corresponds to the slow subsystem. For details of the method, we refer the reader to [9, 10].

In [7, 8] it was shown that this method yields the following effective Hamiltonian $H_{\mathrm{CM,eff}}(\mathbf{R}, \mathbf{P})$ for the center-of-mass motion:

$$\begin{aligned} H_{\mathrm{CM,eff}}(\mathbf{R}, \mathbf{P}) = {}& P^2/[2(m_1 + m_2)] + (e_1 + e_2)\varphi(\mathbf{R}) \\ & - (3n|q|\hbar^2/2)[1/(m_1e_2) + 1/(m_2|e_1|)]F(\mathbf{R}) \cos[\theta(\mathbf{R})]. \end{aligned} \tag{4.19}$$

Here q is the electric quantum number ($q = n_1 - n_2$, where n_1 and n_2 are the parabolic quantum numbers), $\theta(\mathbf{R})$ is the polar angle of the vector $\mathbf{F}(\mathbf{R})$, and the z-axis is chosen along the Runge–Lenz vector $\mathbf{A}$. We note that the eigenvalue of the operator $\mathbf{A}$ is q/n—see e.g. [11–13]. Thus for the motion for any two particles of opposite charges and of significantly different masses, we achieved the pseudoseparation of the relative motion and the center-of-mass in a non-uniform electric field.

For hydrogen atoms one has

$$e_1 = e, \quad e_2 = -e, \qquad \mu = m_e m_p/(m_e + m_p). \tag{4.20}$$

Here m_e and m_p are the electron and proton masses, respectively, and $e > 0$ is the electron charge. Thus equation (4.23) reduces to

$$\begin{aligned} H_{CM,eff}(\mathbf{R}, \mathbf{P}) &= P^2/(2m) - [3n|q|\hbar^2/(2\mu e)]F(\mathbf{R}) \cos[\theta(\mathbf{R})], \\ m &= (m_e + m_p). \end{aligned} \tag{4.21}$$

Now we consider the case where the non-uniform electric field is due to the nearest plasma ion of the positive charge Ze and mass m_i (the nearest to the hydrogen atom) located at the distance $\mathbf{R}$ from the hydrogen atom. In this case, the Hamiltonian from equation (4.25) can be represented as follows:

$$\begin{aligned} H_{CM,eff}(\mathbf{R}, \mathbf{P}) &= P^2/(2m) - (D/R^2)\cos\theta, \\ D &= [3n|q|\hbar^2/(2\mu)]Z, \quad \cos\theta = \mathbf{AR}/\mathrm{AR}. \end{aligned} \tag{4.22}$$

This Hamiltonian mathematically corresponds to a particle of mass m moving in the dipole potential. The motion of the particle can be described classically because the particle is relatively heavy: $m \gg m_e$.

The corresponding classical solution can be found in e.g. [14]. The angular motion and the radial motion can be separated exactly. For the radial motion one obtains

$$m[R(\mathrm{d}R/\mathrm{d}t) + (\mathrm{d}R/\mathrm{d}t)^2] = E_{CM}. \tag{4.23}$$

Here E_{CM} is the total energy of the center-of-mass particle. The exact general solution of equation (4.23) is

$$R(t) = (2E_{CM}t^2/m + 2R_0 v_0 t + R_0^2)^{1/2}, \quad R_0 = R(0), \quad v_0 = (\mathrm{d}R/\mathrm{d}t)_{t=0}. \tag{4.24}$$

In low density plasmas the ion dynamical broadening mechanism controls the Stark broadening of the most intense hydrogen lines—see e.g. [15–21]. In the 'standard theory' [22–24] the relative motion of a hydrogen atom and its nearest perturbing ion was assumed along a straight line. However, the above results demonstrate that the relative motion of these two particles takes place in the dipole potential—$(D/R^2)\cos\theta$. We choose the instant of their closest approach as $t = 0$. Then equation (4.24) reduces to

$$R(t) = (2E_{CM}t^2/m + R_0^2)^{1/2}. \tag{4.25}$$

For the energy E_{CM} one has

$$E_{CM} = P_0^2/(2m) - (D/R_0^2)\cos\theta_0, \qquad P_0 = P(0), \qquad \theta_0 = \theta(0). \tag{4.26}$$

At this point the author [7, 8] proceeded to the reference frame where the perturbing ion is at rest. Then $P_0 = mV_0$, where V_0 is the relative velocity at $t = 0$. Thus, the energy E_{CM} can be rewritten as follows:

$$E_{CM} = mV_0^2/2 - (D/R_0^2)\cos\theta_0. \tag{4.27}$$

Equation (4.25) becomes

$$R(t) = \left\{[V_0^2 - 2D\cos\theta_0/(mR_0^2)]t^2 + R_0^2\right\}^{1/2}. \tag{4.28}$$

The introduction of the effective velocity

$$V_{\rm eff}(R_0, \theta_0) = [V_0^2 - 2D\cos\theta_0/(mR_0^2)]^{1/2} \tag{4.29}$$

formally reduces equation (4.32) to its counterpart for the case of the rectilinear trajectories:

$$R(t) = \left\{[V_{\rm eff}(R_0, \theta_0)]^2 t^2 + R_0^2\right\}^{1/2}. \tag{4.30}$$

To be able to derive the final results analytically, the author [7, 8] uses the so-called impact approximation described in e.g. [23, 24] or [21, 22]. Then the ion dynamical broadening operator Φ_{ab} for a radiative transition between the energy levels a and b of a hydrogen atom can be represented in the form (by analogy with the electron dynamical broadening operator—see e.g. [25]):

$$\Phi_{ab} = -\int dV_0 f(V_0) N_i V_0 \langle\sigma(V_0, \theta_0)\rangle_{\theta_0}, \tag{4.31}$$

$$\sigma(V_0, \theta_0) = \int dR_0 2\pi R_0 [1 - S_a(R_0, V_0, \theta_0) S_b^*(R_0, V_0, \theta_0)]_{\rm ang.av}. \tag{4.32}$$

Here $f(V_0)$ is the distribution of the velocities (usually Maxwellian), N_i is the ion density, $\langle\cdots\rangle\theta_0$ denotes the averaging over the angle θ_0, $[\ldots]_{\rm ang.av}$ stands for the angular average, and the symbol * stands for the complex conjugation.

The shape of the spectral line is a sum of Lorentzians when non-diagonal matrix elements of the operator Φ_{ab} are relatively small. The width $\gamma_{\alpha\beta}$ and shift $\Delta_{\alpha\beta}$ of the Lorentzians are as follows:

$$\gamma_{\alpha\beta} = -{\rm Re}[_{\alpha\beta}(\Phi_{ab})_{\beta\alpha}], \qquad \Delta_{\alpha\beta} = -{\rm Im}[_{\alpha\beta}(\Phi_{ab})_{\beta\alpha}], \tag{4.33}$$

where α and β correspond to the upper and lower sublevels of the levels a and b, respectively.

Here and below, for any operator G, we denote its matrix elements $\langle\alpha|\langle\beta|G|\beta\rangle|\alpha\rangle$ as $_{\alpha\beta}G_{\beta\alpha}$ for brevity.

For the operator σ the author [7, 8] obtained

$$\sigma(R_0, V_0, \theta_0) = \int dR_0 2\pi R[K^2 Q(R_0, V_0, \theta_0)/R_0^2]. \tag{4.34}$$

Here

$$Q(R_0, V_0, \theta_0) = 2\hbar^2/[3\mu^2 V_{\text{eff}}(R_0, \theta_0)^2]Q_0/[1 - 2D\cos\theta_0/(mV_0^2R_0^2)],$$
$$Q_0 = 2Z^2\hbar^2/(3\mu^2 V_0^2), \tag{4.35}$$

and

$$K^2 = K_a^2 + K_{\text{interf}} + K_b^2,\ K_a^2 = \mathbf{r}_a^2/a_B^2,\ K_{\text{interf}} = -2\mathbf{r}_a\mathbf{r}_b^*/a_B^2,$$
$$K_b^2 = \mathbf{r}_b^{*2}/a_B^2,\ a_B = \hbar^2/(\mu e^2). \tag{4.36}$$

In equation (4.36) K_{interf} represents the so-called interference term and a_B is the Bohr radius.

Then the author [7, 8] averaged $1/V_{\text{eff}}(R_0, \theta_0)^2$ in equation (4.35) over the angle θ_0:

$$(1/2)\int_{-1}^{1} d(\cos\theta_0)/[V_0^2 - 2D\cos\theta_0/(mR_0^2)]$$
$$= [R_0^2/(2R_D^2V_0^2)]\ln[(R_0^2 + R_D^2)/(R_0^2 - R_D^2)]. \tag{4.37}$$

Here

$$R_D = [2D/(mV_0^2)]^{1/2}. \tag{4.38}$$

The averaging of the quantity $Q(R_0, \theta_0)$ over θ_0 yields

$$Q(R_0) = \left[Q_0R_0^2/(2R_D^2)\right]\ln\left[\left(R_0^2 + R_D^2\right)/\left(R_0^2 - R_D^2\right)\right]. \tag{4.39}$$

We remind that the quantity Q_0 was defined in equation (4.35).

Figure 4.3 shows the dependence of the ratio Q/Q_0 on R_0 and R_D.

For hydrogen lines other than the Lyman lines, the definition of the quantity D from equation (4.22) has to be modified to allow for the Stark effect of both the upper and lower sublevels. We modify it similarly to the suggestion from [26, 27]:

$$D = 3(n|q| + n'|q'|)Ze^2a_B/4, \tag{4.40}$$

where the quantum numbers without the prime symbol and with the prime relate to the upper and lower sublevels, respectively.

The next step is the averaging over R_0. For the diagonal elements of the cross-section of 'Stark broadening collisions' we have

$$_{\alpha\beta}(\sigma)_{\beta\alpha,D} = \int_{R_{\min}}^{R_{\max}} dR_0 2\pi R_0\left[_{\alpha\beta}(K^2)_{\beta\alpha}Q(R_0)/R_0^2\right] + \int_0^{R_{\min}} dR_0 2\pi R_0 C, \tag{4.41}$$

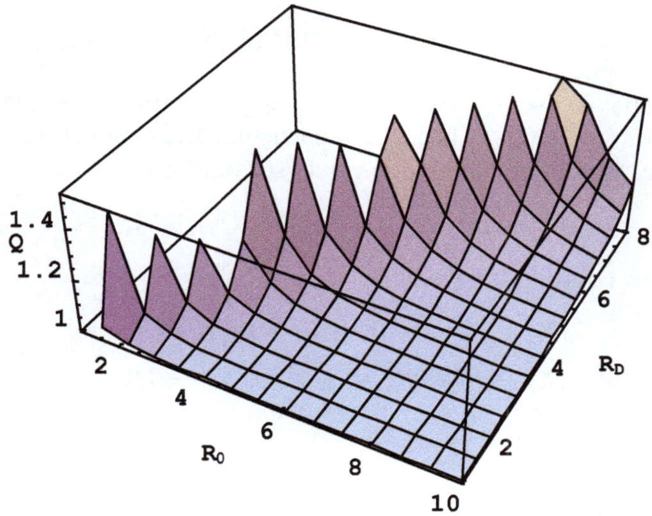

Figure 4.3. Dependence of the quantity Q (in units of the quantity Q_0) on R_0 and R_D (see the text).

where $R_{\min}$ is defined as follows:

$$_{\alpha\beta}(K^2)_{\beta\alpha}Q(R_{\min})/R_{\min}^2 = {}_{\alpha\beta}(K^2)_{\beta\alpha}[Q_0/(2R_D^2)]\ln[(R_{\min}^2 + R_D^2)/(R_{\min}^2 - R_D^2)] = C \quad (4.42)$$

(obviously, $R_{\min} > R_D$). Here and below, the subscript D in $_{\alpha\beta}(\sigma)_{\beta\alpha,D}$ signifies that this cross-section was calculated by allowing for the center-of-mass motion. The constant C in equation (4.42) is the 'strong collision constant' of the standard theory. It is used in the standard theory for preserving the unitarity of the S-matrices:

$$|1 - S_a(R_0, V_0, \theta_0)S_b^*(R_0, V_0, \theta_0)| = C, \qquad C \leqslant 2. \quad (4.43)$$

For example, according to Griem [22, p 43], his choice was $C = 3/2$. Additional details can be found in [28].

The Debye radius is chosen as the upper cutoff $R_{\max}$,

$$R_{\max} = R_D = [T/(4\pi e^2 N_e)]^{1/2}. \quad (4.44)$$

The integration over R_0 can be performed analytically. Then, using the expression from equation (4.42) for the strong collision constant C, we obtain

$$\begin{aligned} {}_{\alpha\beta}(\sigma)_{\beta\alpha,D} = 2\pi{}_{\alpha\beta}(K^2)_{\beta\alpha}Q_0\Big\{ &\ln\Big[\big(R_{\max}^4 - R_D{}^4\big)^{1/4}/\big(R_{\min}^4 - R_D{}^4\big)^{1/4}\Big] \\ &+ \big[R_{\max}^2/(4R_D{}^2)\big]\ln\Big[\big(R_{\max}^2 + R_D{}^2\big)/\big(R_{\max}^2 - R_D{}^2\big)\Big]\Big\}. \end{aligned} \quad (4.45)$$

The weak and strong collisions in equation (4.41) are separated by the boundary $R_{\min}$. It can be determined by solving equation (4.42) with respect to $R_{\min}$:

$$R_{\min} = R_D\left\{\left[\exp\left(2CR_D{}^2/_{\alpha\beta}(K^2)_{\beta\alpha}Q_0\right) + 1\right]\Big/\left[\exp\left(2CR_D{}^2/_{\alpha\beta}(K^2)_{\beta\alpha}Q_0\right) - 1\right]\right\}^{1/2}. \quad (4.46)$$

The next step is to average several of the above quantities over Stark sublevels of the upper and lower levels. Then, for a particular hydrogen spectral line, each of these quantities would have a unique value, starting from

$$[_{\alpha\beta}(K^2)_{\beta\alpha}]_{\rm av}{}^{1/2} = [_{\alpha}(K_{\rm a}{}^2)_{\alpha}{}^{1/2} - {}_{\beta}(K_{\rm b}{}^2)_{\beta}{}^{1/2}]_{\rm av}. \quad (4.47)$$

In the parabolic coordinates, we have (see e.g. [12, 29])

$$\begin{aligned} {}_{\alpha}(K_{\rm a}^2)_{\alpha} &= (9/8)n^2(n^2 + q^2 - m^2 - 1), \\ {}_{\beta}(K_{\rm b}^2)_{\beta} &= (9/8)n'^2(n'^2 + q'^2 - m'^2 - 1)], \end{aligned} \quad (4.48)$$

so that

$$\left[_{\alpha\beta}(K^2)_{\beta\alpha}\right]_{\rm av} = (9/8)(n^2 - n'^2). \quad (4.49)$$

We denote

$$R_{\rm WA}(C) = \left\{\left[_{\alpha\beta}(K^2)_{\beta\alpha}\right]_{\rm av}Q_0/C\right\}^{1/2} = (3/C)^{1/2}(n^2 - n'^2)\hbar Z/(2\mu V_0). \quad (4.50)$$

This quantity is the Weisskopf radius, more accurately defined than in the standard theory [22]—see appendix B of [8]. Therefore, here and below, we use the superscript 'A' to mean 'accurate'.

The averaging of the quantity D yields

$$\langle D\rangle_{\rm av} = (n^2 + n'^2)Ze^2a_{\rm B}/4. \quad (4.51)$$

By using $\langle D\rangle_{\rm av}$ from equation (4.51) in equation (4.38) we obtain

$$\langle R_{\rm D}\rangle_{\rm av} = [(n^2 + n'^2)Z/2]^{1/2}\hbar/(\mu V_0). \quad (4.52)$$

Thus

$$\langle R_{\min}\rangle_{\rm av} = \langle R_{\rm D}\rangle_{\rm av}\left\{[\exp(2\langle R_{\rm D}\rangle_{\rm av}^2/R_{\rm WA}(C)^2) + 1]/[\exp(2\langle R_{\rm D}\rangle_{\rm av}^2/R_{\rm WA}(C)^2) - 1]\right\}^{1/2}. \quad (4.53)$$

Now in equation (4.45) we substitute $R_{\rm D}$ by $\langle R_{\rm D}\rangle_{\rm av}$ and $R_{\min}$ by $\langle R_{\min}\rangle_{\rm av}$. Then we define the following dimensionless parameters

$$\begin{aligned} w = \langle R_{\rm D}\rangle_{\rm av}/R_{\max}, \quad b &= \langle R_{\rm D}\rangle_{\rm av}/R_{\rm WA}(C) \\ &= [2C/(3Z)]^{1/2}(n^2 + n'^2)^{1/2}/(n^2 - n'^2). \end{aligned} \quad (4.54)$$

As a result we obtain

$$\begin{aligned} {}_{\alpha\beta}(\sigma)_{\beta\alpha,\rm AD} = 2\pi_{\alpha\beta}(K^2)_{\beta\alpha}Q_0\Big\{ &\ln[(\exp(2b^2) - 1)^{1/2}(1/w^4 - 1)^{1/4}/2^{1/2}] \\ &- b^2/2 + [1/(4w^2)]\ln[(1 + w^2)/(1 - w^2)]\Big\}, \end{aligned} \quad (4.55)$$

where

$$ {}_{\alpha\beta}(K^2)_{\beta\alpha} = (9/8)[n^2(n^2 + q^2 - m^2 - 1) - 4nqn'q' + n'^2(n'^2 + q'^2 - m'^2 - 1)]. \tag{4.56} $$

Equation (4.54) shows that the ratio $Z^{1/2}b/C^{1/2}$ is just a combination of the principal quantum numbers n and n'. It does not depend on the temperature T and the electron density N_e.

In distinction, the parameter $w = \langle R_D\rangle_{av}/R_{max}$ from equation (4.54) depends on the plasma parameters:

$$ \begin{aligned} w &= [2e\hbar/(\mu T)][(n^2 + n'^2)Zm_r N_e]^{1/2} \\ &= 8.99 \times 10^{-10}\left[(n^2 + n'^2)ZN_e m_r/m_p\right]^{1/2}/T, \end{aligned} \tag{4.57} $$

where

$$ m_r = (m_e + m_p)m_i/(m_e + m_p + m_i). \tag{4.58} $$

In the practical formula (4.57) the temperature T is in electronvolts and the electron density N_e is in cubic centimetres. For the Lyman lines the expression for w should read

$$ w = [e\hbar n/(\mu T)](2m_r ZN_e)^{1/2} = 1.27 \times 10^{-9}n[ZN_e m_r/m_p]^{1/2}/T. \tag{4.59} $$

It is instructive to analyze the ratio of the cross-section ${}_{\alpha\beta}(\sigma)_{\beta\alpha,\mathrm{A,D}}$ to the corresponding cross-section ${}_{\alpha\beta}(\sigma)_{\beta\alpha,\mathrm{G}}$ from the conventional theory by Griem [22] (which is practically the same as the ratio of widths $\gamma_{\alpha\beta,\mathrm{A,D}}/\gamma_{\alpha\beta,\mathrm{G}}$):

$$ \begin{aligned} \text{ratio} &= {}_{\alpha\beta}(\sigma)_{\beta\alpha,\mathrm{A,D}}/{}_{\alpha\beta}(\sigma)_{\beta\alpha,\mathrm{G}} = \gamma_{\alpha\beta,\mathrm{A,D}}/\gamma_{\alpha\beta,\mathrm{G}} \\ &= \{\ln[(\exp(2b^2) - 1)^{1/2}(1/w^4 - 1)^{1/4}/2^{1/2}] - b^2/2 \\ &\quad + [1/(4w^2)]\ln[(1 + w^2)/(1 - w^2)]\}/\{\ln[b/(wC^{1/2})] + 0.356\}. \end{aligned} \tag{4.60} $$

The matrix element ${}_{\alpha\beta}(W^2)_{\beta\alpha}$ does not enter this ratio. Therefore this ratio is a universal function of just two dimensionless parameters w and b. In this form equation (4.60) can be utilized for any set of the five parameters N_e, T, n, n', and C.

Below we present numerical examples for some laboratory and astrophysical plasmas. In both the edges of magnetic fusion machines (for example tokamaks) and the atmospheres of flare stars, the typical plasma parameters are $N_e = (10^{14}\text{–}10^{15})$ cm^{-3} and the temperature is one or a few electonvolts (see e.g. [30–32]), so that for the Stark profiles of the lines, such as Lyman-alpha, Lyman-beta, H-alpha, and so on, the ion dynamical broadening can be the dominant mechanism (see e.g. [15–21]). For a hydrogen plasma at $N_e = 5 \times 10^{14}$ cm^{-3} and $T = 1$ eV, for the H-alpha line, the ratio from equation (4.60) yields 1.19 for $C = 2$ and 1.13 for $C = 3/2$. Figure 4.4 shows this ratio versus the electron density N_e at $T = 1$ eV for the H-alpha line: for $C = 2$ (solid line) and for $C = 3/2$ (dashed line). Figure 4.4 demonstrates that the ion dynamical Stark width of this line increases by up to 15%–20% when the center-of-mass motion is taken into account.

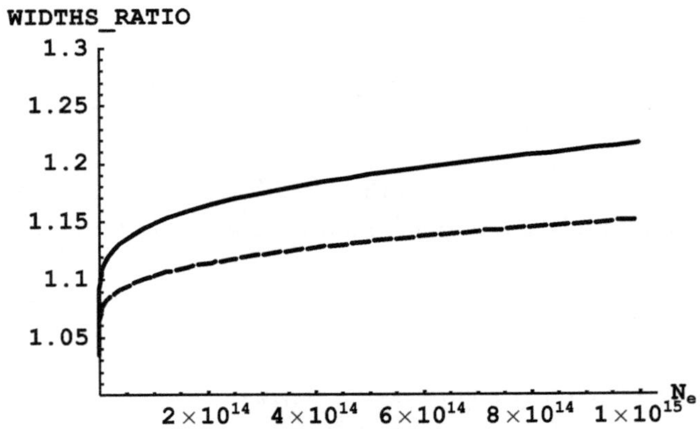

Figure 4.4. The ratio of the ion dynamical Stark width with the allowance for the center-of-mass motion to the ion dynamical Stark width from the conventional theory [22] versus the electron density N_e (cm^{-3}) for the H_α line emitted from a hydrogen plasma at $T = 1$ eV for $C = 2$ (solid line) and for $C = 3/2$ (dashed line). The plasma parameters correspond to the edge plasmas of magnetic fusion machines and to the atmospheres of flare stars. (Reproduced with permission from [7]. Copyright 2017 E Oks.)

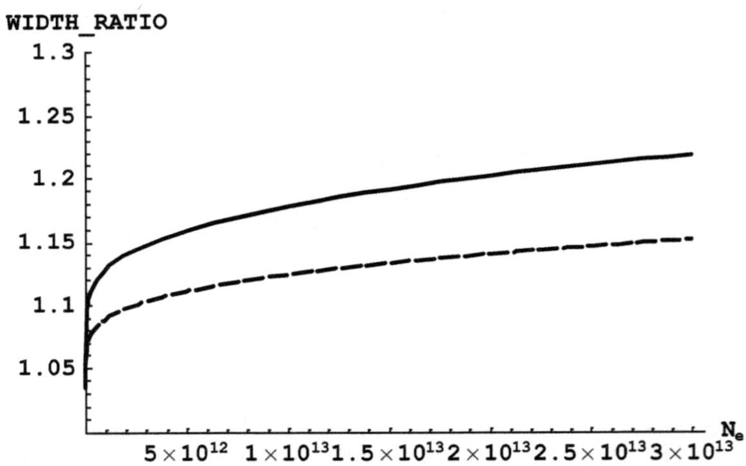

Figure 4.5. The ratio of the ion dynamical Stark width with the allowance for the center-of-mass motion to the ion dynamical Stark width from the conventional theory [22] versus the electron density N_e (cm^{-3}) for the H_α line emitted from a hydrogen plasma at $T = 0.17$ eV for $C = 2$ (solid line) and for $C = 3/2$ (dashed line). Plasma parameters correspond to radiofrequency discharges. (Reproduced with permission from [8]. Copyright 2018 E Oks.)

Another example is the plasmas of radiofrequency discharges [33–35]. These are characterized (in experiments [33, 34]) by $N_e = 1.2 \times 10^{13}$ cm^{-3} and $T = (0.16–0.17)$ eV. The for the H-alpha line, for the ratio from equation (4.60), we find: 1.18 for $C = 2$ and 1.13 for $C = 3/2$. Figure 4.5 illustrates this ratio versus the electron density N_e at $T = 0.17$ eV for $C = 2$ (solid line) and for $C = 3/2$ (dashed line). Thus, for plasmas of radiofrequency discharges, the ion dynamical Stark width of this line also increases by up to 15%–20% when the center-of-mass motion is taken into account.

References

[1] Schmelcher P and Cederbaum L S 1995 *Phys Rev. Lett.* **74** 662
[2] Schmelcher P and Cederbaum L S 1993 *Phys. Rev.* A **47** 2634
[3] Vincke M and Baye D 1988 *J. Phys. B: At. Mol. Opt. Phys.* **21** 2407
[4] Schmelcher P and Cederbaum L S 1993 *Chem. Phys. Letters* **208** 548
[5] Schmelcher P and Cederbaum L S 1991 *Phys. Rev.* A **43** 287
[6] Kotkin G L and Serbo V G 1971 *Collection of Problems in Classical Mechanics* (Oxford: Pergamon) problem 2.22
[7] Oks E 2017 *Int. Rev. At. Mol. Phy.* **8** 41
[8] Oks E 2018 *J. Phys. Commun.* **2** 045005
[9] Oks E 2020 *Analytical Advances in Quantum and Celestial Mechanics: Separating Rapid and Slow Subsystems* (Bristol: IOP Publishing) ch 10
[10] Galitski V, Karnakov B, Kogan V and Galitski V Jr 2013 *Exploring Quantum Mechanics* (Oxford: Oxford University Press) p 363
[11] Landau L D and Lifshitz E M 1965 *Quantum Mechanics* (Oxford: Pergamon)
[12] Lisitsa V S 1977 *Sov. Phys. Usp.* **122** 603
[13] Lisitsa V S 1994 *Atoms in Plasmas* (Berlin: Springer) p 13
[14] Fox K 1968 *J. Phys. A: Gen. Phys.* **1** 124
[15] Abramov V A and Lisitsa V S 1977 *Sov. J. Plasma Phys.* **3** 451
[16] Seidel J 1979 *Z. Naturforsch.* **34a** 1385
[17] Stehle C and Feautrier N 1984 *J. Phys.* B **17** 1477
[18] Derevianko A and Oks E 1994 *Phys. Rev. Lett.* **73** 2059
[19] Derevianko A and Oks E 1995 *J. Quant. Spectr. Rad. Transfer* **54** 137
[20] Derevianko A and Oks E 1997 *Rev. Sci. Instrum.* **68** 998
[21] Oks E 2006 *Stark Broadening of Hydrogen and Hydrogenlike Spectral Lines in Plasmas: The Physical Insight* (Oxford: Alpha Science International)
[22] Griem H R 1974 *Spectral Line Broadening by Plasmas* (New York: Academic)
[23] Baranger M 1958 *Phys. Rev.* **111** 494
[24] Kepple P and Griem H R 1968 *Phys. Rev.* **173** 317
[25] Kolb A C and Griem H R 1958 *Phys. Rev.* **111** 514
[26] Nienhuis G 1973 *Physica* **66** 245
[27] Szudy J and Baylis W E 1976 *Can. J. Phys.* **54** 2287
[28] Oks E 2015 *J. Quant. Spectr. Rad. Transfer* **152** 74
[29] Sholin G V, Demura A V and Lisitsa V S 1973 *Sov. Phys. JETP* **37** 1057
[30] Pospieszczyk A 2005 *Phys. Scripta* **2005** T119
[31] Gershberg R E 2005 *Solar-Type Activity in Main-Sequence Stars* (Berlin: Springer)
[32] Oks E and Gershberg R E 2016 *Astrophys. J.* **819** 16
[33] Bengston R D, Tannich J D and Kepple P 1970 *Phys. Rev.* A **1** 532
[34] Bengtson R D and Chester G R 1972 *Astrophys. J.* **178** 565
[35] Himmel G 1976 *J. Quant. Spectrosc. Rad. Transfer* **16** 529

Chapter 5

Ionization of hydrogen atoms by a low-frequency laser field of arbitrary strength: no 'local suppression'

There are two regimes of the ionization of atoms, including hydrogen atoms: multiphoton ionization and tunneling ionization—see e.g. [1–3]. They are distinguished by large and small values of the following parameter [4]:

$$\gamma = \omega(2E_i)^{1/2}/F. \tag{5.1}$$

In equation (5.1) F and ω are the characteristic strength and the frequency of the oscillatory electric field (OEF), respectively. The typical assumptions are $F \ll F_a$ and $\hbar\omega \ll E_i$. Here, F_a is the atomic electric field and E_i is the binding energy of the outer atomic electron. Below atomic units are used (unless specified to the contrary).

The multiphoton ionization is characterized by $\gamma \gg 1$. The tunneling ionization is characterized by $\gamma \ll 1$. For hydrogenic atoms or ions, the ionization rate for $\gamma \ll 1$ is

$$W(F) = 4[2U(F)]^{5/2}F^{-1}\exp\left\{-(2/3)[2U(F)]^{3/2}/F\right\}. \tag{5.2}$$

Here $U(F)$ is the absolute value of the energy of the atomic level. For experimental results we refer to [5–7].

For a strong field F equation (5.2) is invalid. Thus in [8] the authors put forward the idea of using, instead of the true potential, some effective potential such that for the latter the Schrödinger equation can be solved exactly. Following this idea, Kulyagin and Taranukhin [9] (hereafter, KT) calculated the ionization rate $W(F)$ of hydrogen atoms. Specifically, they calculated this value for the ground state in the case of a linear polarization of the laser field. For the substitute-barrier, KT chose some effective parabolic barrier and used the known analytical result (e.g. [10]) for the transmission coefficient. As a result, they claimed that the dependence of the transmission coefficient on F has the second minimum (the first being at $F = 0$) at $F = 0.2$ a.u. Previously, this kind of a *stabilization* (with respect to the ionization)

doi:10.1088/978-0-7503-3679-6ch5

was considered to be impossible for the tunneling ionization regime (in the frame of which KT's paper [9] was written) and to be possible only for the multiphoton ionization [11–13].

In [14] the authors showed that KT made principal errors and that there is no stabilization. Below is an overview of [14].

The central part is the differential equation for the 'η-part' of the wave function:

$$d^2\chi/d\eta^2 + [E(F)/2 + \beta_2(F)/\eta + F\eta/4 - (m^2 - 1)/(4\eta^2)]\chi = 0. \tag{5.3}$$

In equation (5.3) $\eta = r - z$ is one of the parabolic coordinates (here r is the absolute value of the radius-vector, the z-axis is parallel to the laser field), $E(F)$ is the field-dependent energy, and $\beta_2(F)$ is the separation constant. Equation (5.3) is mathematically equivalent to the one-dimensional Schrödinger equation with the following features. The 'total energy' is $E(F)/4$; the effective potential energy is

$$U_2(\eta) = -\beta_2/(2\eta) - F\eta/8 + (m^2 - 1)/(8\eta^2). \tag{5.4}$$

For the ground state ($m = 0$) equation (5.4) reduces to

$$U_2(\eta, F) = -\beta_2(F)/(2\eta) - F\eta/8 - 1/(8\eta^2). \tag{5.5}$$

One of the errors by KT was the omission of the last term in $U_2(\eta)$.

Here is the succession of steps taken by Gavrilenko and Oks [14] according to Miller–Good's method from [8]. The classical turning points can be found from the equation

$$U_2(\eta, F) - E(F)/4 = 0. \tag{5.6}$$

We denote the absolute value of the hydrogen energy level as

$$U(F) = -E(F). \tag{5.7}$$

The argument F of the functions U, U_2, and β_2 is omitted below for brevity.

The explicit form of equation (5.10) is

$$\begin{aligned} U_2(\eta) + U/4 &= -[F/(8\eta^2)](\eta^3 - 2U\eta^2/F + 4\beta_2\eta/F + 1/F) \\ &= [F/(8\eta^2)](b - \eta)(\eta - a)(\eta + c) = 0. \end{aligned} \tag{5.8}$$

This is a cubic equation and it has one negative (real) root denoted as $c(F)$, so that $c(F) > 0$. Depending on the laser field amplitude F, the two other roots $b(F)$ and $a(F)$ either are real and positive, or are complex numbers (mutually conjugated) having a positive real part. Gavrilenko and Oks [14] assumed that the real part of $a(F)$ is smaller than (or equal to) the real part of $b(F)$.

Gavrilenko and Oks [14] denoted the left side of equation (5.8) by $J(F)$

$$J(F) = (F/8)^{1/2}\,\mathrm{Int}(F), \quad \mathrm{Int}(F) = \int_a^b (dx/x)[(b - x)(x - a)(x + c)]^{1/2}, \tag{5.9}$$

and introduced the following notations:

$$d = [(a + c)/(b + c)]^{1/2}, \quad j = [a/(b + c)]^{1/2}. \tag{5.10}$$

For Int(*F*) they obtained the following analytical result:

$$\begin{aligned} \mathrm{Int} = &- i(2/3)(a + c)^{-1/2}\Big\{[b(a + c) + (a^2 - c^2)][\mathbf{E}(1/d^2) - \mathbf{E}(\mathrm{arcsin}d, 1/d^2)] \\ &- [b(2c - a) + a(a + c)][\mathbf{K}(1/d^2) - \mathbf{F}(\mathrm{arcsin}d, 1/d^2)] \\ &- 3ab\Big[\mathbf{\Pi}\big(1/(d^2 - j^2), 1/d^2\big) - \mathbf{\Pi}\big(1/(d^2 - j^2), \mathrm{arcsin}d, 1/d^2\big)\Big]\Big\}. \end{aligned} \tag{5.11}$$

Here $\mathbf{K}(m)$, $\mathbf{E}(m)$, and $\mathbf{\Pi}(n,m)$ are the complete elliptic integrals of the first, second, and third kinds, respectively; and $\mathbf{F}(\varphi,m)$, $\mathbf{E}(\varphi,m)$, and $\mathbf{\Pi}(n,\varphi,m)$ are elliptic integrals of the first, second, and third kinds, respectively.

The well-known (e.g. from [10]) exact expression for the transmission coefficient *D*(*F*) of the parabolic barrier is

$$D = 1/[1 + \exp(-2\pi\alpha)], \tag{5.12}$$

where

$$\alpha(F) = -(b - a)[(U/4 - U_{0p})/8]^{1/2}. \tag{5.13}$$

In appendix A of this book, we present the derivation of equation (5.13) and correct yet another erroneous result of KT, namely in the expression for $\alpha(F)$.

Figure 5.1 illustrates the dependence of the transmission coefficient *D* on the parameter α.

By combining equations (5.4), (5.12), and (5.13) Gavrilenko and Oks [14] derived the expression for the transmission coefficient as follows:

$$D(F) = 1/\{1 + \exp[2^{3/2}J(F)]\} = 1/\{1 + \exp[F^{1/2}\,\mathrm{Int}(F)]\}, \tag{5.14}$$

where *J*(*F*) is given by equation (5.9).

The next step was the calculation of the ionization rate *W*(*F*):

$$W(F) = S(F)D(F). \tag{5.15}$$

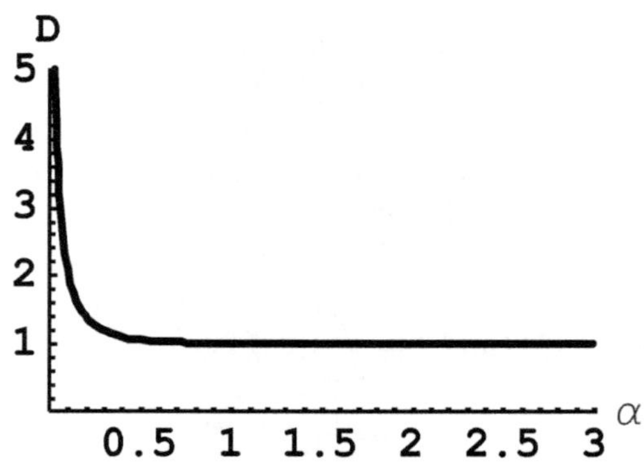

Figure 5.1. Dependence of the transmission coefficient *D* on the parameter α (see equations (5.12) and (5.13)).

Here $S(F)$ is the input flow probability, which can be found by the transition to the weak field limit ($F \ll F_a$) as shown below.

The weak field limit of Int(F)—denoted Int$_{as}$(F)—can be presented in the form (see appendix B of this book)

$$\begin{aligned}\mathrm{Int}_{as}(F) = (2/3)(2U/F)^{3/2} - (4/3)(2/F)^{1/2}(1/2 + \beta_2{}^2/U)^{1/2} \\ \times \ln\{[8U^2/F - 12\beta_2 + 4(\beta_2{}^2 + U/2)^{1/2}]/(\beta_2{}^2 + U/2)^{1/2}\} \\ + (2/F)^{1/2}[(1/3 + \ln 2)(1/2 + \beta_2{}^2/U)^{1/2} - (1 + \ln 2)\beta_2].\end{aligned} \tag{5.16}$$

The weak field limit of $J(F)$—denoted $J_{as}(F)$—can be obtained by multiplying equation (5.16) by $(F/8)^{1/2}$ (see equation (5.7)):

$$\begin{aligned}J_{as}(F) = 2U^{3/2}/(3F) - (2/3)(1/2 + \beta_2{}^2/U)^{1/2} \\ \times \ln\{[8U^2/F - 12\beta_2 + 4(\beta_2{}^2 + U/2)^{1/2}]/(\beta_2{}^2 + U/2)^{1/2}\} \\ + (1/2)[(1/3 + \ln 2)(1/2 + \beta_2{}^2/U)^{1/2} - (1 + \ln 2)\beta_2].\end{aligned} \tag{5.17}$$

On substituting equation (5.17) in equation (5.14) and utilizing equation (5.2) for the ionization rate $W(F)$ in the weak field limit, Gavrilenko and Oks [14] obtained the input flow probability:

$$\begin{aligned}S(F) = [4(2U)^{5/2}/F]\exp\{(1/2)[(1/3 + \ln 2)(1/2 + \beta_2{}^2/U)^{1/2} - (1 + \ln 2)\beta_2]\} \\ \times [(\beta_2{}^2 + U/2)^{1/2}/[8U^2/F - 12\beta_2 + 4(\beta_2{}^2 + U/2)^{1/2}]]^{\gamma}, \\ \gamma = (4/3)\left(1 + 2\beta_2{}^2/U\right)^{1/2}.\end{aligned} \tag{5.18}$$

To summarize, the tunneling ionization rate $W(F)$ can be presented in the following form *both for weak and strong fields*:

$$W(F) = S(F)D[J(F)], \quad J(F) = (F/8)^{1/2}\,\mathrm{Int}(F), \tag{5.19}$$

where the expression for Int(F) is presented in equation (5.9).

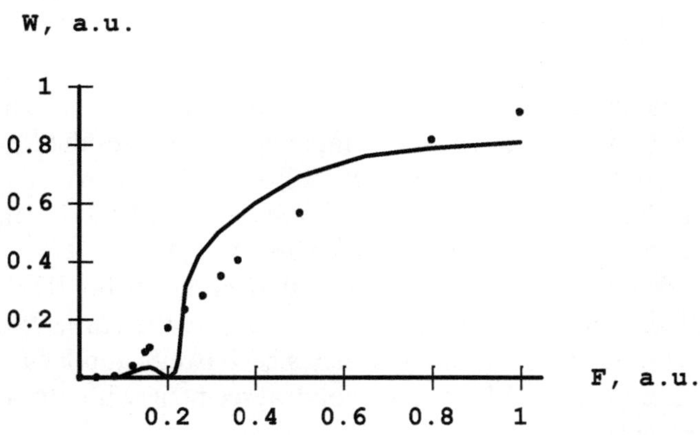

Figure 5.2. Dependence of the ionization rate W of hydrogen atoms in the ground state on the laser field F (W and F are in atomic units) [14]. The dots show our analytical results and the solid line the analytical results of Kulyagin and Taranukhin [9].

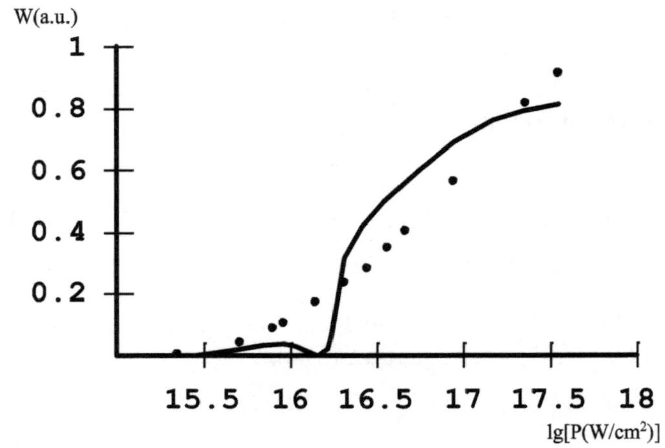

Figure 5.3. Dependence of the ionization rate W (in atomic units) of hydrogen atoms in the ground state on the laser power density P (in W cm^{-2}) [14]. The dots show our analytical results and the solid line the analytical results of Kulyagin and Taranukhin [9].

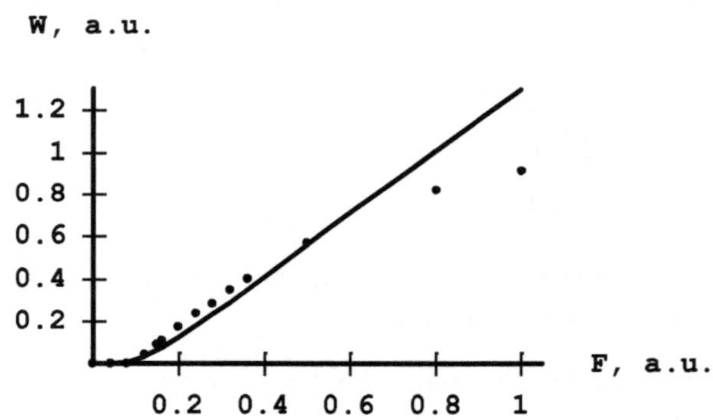

Figure 5.4. Dependence of the ionization rate W of hydrogen atoms in the ground state on the laser field F (W and F are in atomic units) [14]. The dots show our analytical results and the solid line shows fully numerical simulations by Farrelly and Reinhardt [15].

For further calculations, we used the results from [15] for $U(F)$ and from [16] for $\beta_2(F)$. Figures 5.2 and 5.3 present the comparison of our results for the ionization rate W with the corresponding KT results. It is seen that the true dependence of W on the laser field F or on the laser power P is *monotonic*. The difference with the KT results is caused by the above-listed errors made by KT.

Figure 5.4 is devoted to comparing our analytical results for $W(F)$ with quantum fully numerical simulations from [15]. (We note that the earlier numerical calculations from [17] were extended to larger fields by Farrelly and Reinhardt [15].) It is seen that our analytical results are in good agreement with the fully numerical calculation from [15].

References

[1] Wuilleumier F J and Meyer M 2006 *J. Phys. B: At. Mol. Opt. Phys.* **39** R425
[2] Delone N B and Krainov V P 1998 *Phys. Usp.* **41** 469
[3] Popov V S 2004 *Phys. Usp.* **47** 855
[4] Keldysh L V 1965 *Sov. Phys. JETP* **20** 1307
[5] Chin S L, Yergeau F and Lavigne P 1985 *J. Phys. B: At. Mol. Opt. Phys.* **18** L213
[6] Yergeau F, Chin S L and Lavigne P 1987 *J. Phys. B: At. Mol. Opt. Phys.* **20** 723
[7] Corcum P B, Burnett N H and Brunel F 1989 *Phys. Rev. Lett.* **62** 1259
[8] Miller S C Jr and Good R H 1953 *Phys. Rev.* **91** 174
[9] Kulyagin R V and Taranukhin V D 1993 *Laser Phys.* **3** 644
[10] Landau L D and Lifshitz E M 1965 *Quantum Mechanics: Non-Relativistic Theory* (Oxford: Pergamon)
[11] Su Q and Eberly J H 1990 *J. Opt. Soc. Am.* B **7** 564
[12] Dorr M, Potvliege R M and Shakeshaft R 1990 *Phys. Rev. Lett.* **64** 2003
[13] Fedorov M V, Ivanov M Y and Lerner P B 1990 *J. Phys. B: At. Mol. Opt. Phys.* **23** 2505
[14] Gavrilenko V P and Oks E 2011 *Can. J. Phys.* **89** 849
[15] Farrelly D and Reinhardt W P 1983 *J. Phys. B: At. Mol. Opt. Phys.* **16** 2103
[16] Kondratovich V D and Ostrovsky V N 1980 *Zh. Exper. Teor. Fiz.* **79** 395
[17] Hehenberger M, McIntosh H V and Brändas E 1974 *Phys. Rev.* A **10** 1494

Chapter 6

Generalized dynamics of a spherical harmonic oscillator

A sample of the background literature on the subject can be found in [1–10] and references therein. Here we follow [11].

The system under consideration is a charged spherical harmonic oscillator (SHO). The energy E is a conserved quantity. Also, the quantities, whose conservation follows from the geometrical symmetry of the system, are the components of the angular momentum vector $\mathbf{M}$.

In addition, from the algebraic symmetry of the system follows the conservation of the quantities

$$I_{mn} = p_m p_n/\mu + kx_m x_n, \quad m = 1, 2, 3, \qquad n = 1, 2, 3. \tag{6.1}$$

Here μ is the mass of the SHO and p_m and x_m are the Cartesian components of the momentum $\mathbf{p}$ and of the radius-vector $\mathbf{r}$, respectively. Only six of them are independent because $I_{nm} = I_{mn}$. The quantities I_{mn} 'commute' with each other. The unperturbed Hamiltonian H can be written in the form

$$H = (I_{11} + I_{22} + I_{33})/2. \tag{6.2}$$

The orbit is limited to a plane. The x_3-axis (the z-axis) is perpendicular to the orbital plane, by our choice. The dynamical variables are x_1, p_1, x_2, and p_2.

The generalized Hamiltonian (according to the application of Dirac generalized Hamiltonian dynamics (GHD) to atomic systems developed in [8]) is

$$\begin{aligned} H_g = (I_{11} + I_{22})/2 + B_{11}(E)(I_{11} - I_{11,0}) + B_{22}(E)(I_{22} - I_{22,0}) \\ + B_{12}(E)(I_{12} - I_{12,0}). \end{aligned} \tag{6.3}$$

Here B_{11}, B_{22}, and B_{12} are yet unknown coefficients and $I_{mn,0}$ are the values of these conserved quantities in the particular state of the system.

doi:10.1088/978-0-7503-3679-6ch6

The consistency conditions (see [8] for the explanation) are such that the Poisson bracket of any two of them vanishes. Therefore, the consistency conditions from equation (6.10) in this case reduce to equating to zero the Poisson brackets of the components of the angular momentum **M** with the second term on the right-hand side of equation (6.21):

$$[M_i,\, B_{mn}I_{mn}] = \left[e_{ijq}x_jp_q,\, B_{mn}(p_mp_n/\mu + kx_mx_n)\right] = 0. \tag{6.4}$$

Here we use the Levi-Civita symbol e_{ijq}.

Now we calculate the Poisson brackets from equation (6.4) and then substitute I_{mn} by $I_{mn,0}$ (the GHD requirement) to obtain the following relations:

$$B_{22}I_{12,0} = B_{12}I_{22,0}, \tag{6.5}$$

$$B_{12}I_{11,0} = B_{11}I_{12,0}. \tag{6.6}$$

The quantities I_{11} and I_{22} are non-negatively defined (see equation (6.1)). Let us assume (for definiteness) that $I_{11,0} > 0$. Then from equations (6.5) and (6.6) follows

$$B_{12} = B_{11}I_{12,0}/I_{11,0}, \qquad B_{22} = B_{11}I_{22,0}/I_{11,0}. \tag{6.7}$$

After the above step, there remains only one unknown coefficient in H_g:

$$\begin{aligned} H_g = (I_{11} + I_{22})/2 + B_{11}(E)\{&(I_{11} - I_{11,0}) + I_{22,0}/I_{11,0}(I_{22} - I_{22,0}) \\ &+ I_{12,0}/I_{11,0}(I_{12} - I_{12,0})\}. \end{aligned} \tag{6.8}$$

The Hamilton equations of the motion are

$$\begin{aligned} \mathrm{d}x_1/\mathrm{d}t &= \{(1 + 2B_{11})p_1 + (B_{11}I_{12,0}/I_{11,0})p_2, \\ \mathrm{d}x_2/\mathrm{d}t &= (1 + 2B_{11}I_{22,0}/I_{11,0})p_2 + (B_{11}I_{12,0}/I_{11,0})p_1\}/\mu, \end{aligned} \tag{6.9}$$

$$\begin{aligned} \mathrm{d}p_1/\mathrm{d}t &= -\,k\{(1 + 2B_{11})x_1 + (B_{11}I_{12,0}/I_{11,0})x_2, \\ \mathrm{d}p_2/\mathrm{d}t &= (1 + 2B_{11}I_{22,0}/I_{11,0})p_2 + (B_{11}I_{12,0}/I_{11,0})x_1\}. \end{aligned} \tag{6.10}$$

The next relations are obtained by taking the time derivative of equation (6.9) and then by the substitution of equation (6.10) into the result

$$\begin{aligned} \mathrm{d}^2x_1/\mathrm{d}t^2 = -\,\omega_0{}^2\{&[(1 + 2B_{11})^2 + (B_{11}I_{12,0}/I_{11,0})^2]x_1 \\ &+ 2(B_{11}I_{12,0}/I_{11,0})[1 + B_{11}(1 + I_{22,0}/I_{11,0})]x_2\}, \\ \mathrm{d}^2x_2/\mathrm{d}t^2 = -\,\omega_0{}^2\{&[(1 + 2B_{11}I_{22,0}/I_{11,0})^2 + (B_{11}I_{12,0}/I_{11,0})^2]x_2 \\ &+ 2(B_{11}I_{12,0}/I_{11,0})[1 + B_{11}(1 + I_{22,0}/I_{11,0})]x_2\}. \end{aligned} \tag{6.11}$$

Here

$$\omega_0 = (k/\mu)^{1/2} \tag{6.12}$$

is the 'unperturbed' frequency of the SHO.

A solution of the system (6.11) is sought in the form

$$x_1 = \exp(i\omega_g t), \quad x_2 = \alpha \exp(i\omega_g t), \qquad \alpha = \text{const}, \tag{6.13}$$

where ω_g is the generalized frequency (yet unknown) of the SHO.

The substitution of x_1 and x_2 from equation (6.13) into the first equation in expression (6.11) leads to

$$\begin{aligned}\omega^2/\omega_0^2 = {} & (1 + 2B_{11})^2 + (B_{11}I_{12,0}/I_{11,0})^2 \\ & + 2\alpha(B_{11}I_{12,0}/I_{11,0})[1 + 2B_{11}(1 + I_{22,0}/I_{11,0})].\end{aligned} \tag{6.14}$$

The substitution of x_1 and x_2 from equation (6.13) into the second equation in expression (6.29) leads to

$$\begin{aligned}\omega_g^2/\omega_0^2 = {} & (1 + 2B_{11}I_{22,0}/I_{11,0})^2 + (B_{11}I_{12,0}/I_{11,0})^2 \\ & + (2/\alpha)(B_{11}I_{12,0}/I_{11,0})[1 + 2B_{11}(1 + I_{22,0}/I_{11,0})].\end{aligned} \tag{6.15}$$

Equations (6.14) and (6.15) have the same left-hand sides. For the compatibility of equations (6.14) and (6.15), their right-hand sides should be equal to each other as well. The result obtained from this outcome is that the parameter α must be a root of the following quadratic equation:

$$\alpha^2 - 2\gamma\alpha - 1 = 0, \qquad \gamma = (I_{22,0} - I_{11,0})/I_{12,0}. \tag{6.16}$$

Equation (6.16) has two solutions:

$$\alpha_{\pm} = \gamma \pm (\gamma^2 + 1)^{1/2}. \tag{6.17}$$

Clearly $\alpha_+ > 0$ while $\alpha_- < 0$. There are two opposite directions of the revolution along the orbit (see equation (6.13)) and the above two solutions α_+ and α_- correspond to the two directions of the revolution.

In a more explicit form, equation (6.15) is

$$\omega_g^2/\omega_0^2 = (4 + 2\alpha_{\pm}\varepsilon\delta + \delta^2)B_{11}^2 + 2(2 + \alpha_{\pm}\delta)B_{11} + 1. \tag{6.18}$$

In equation (6.18) we used the following temporary notations:

$$\varepsilon = (1 + I_{22,0}/I_{11,0}), \qquad \delta = I_{12,0}/I_{11,0}. \tag{6.19}$$

From equation (6.17), we obtain

$$4 + 2\alpha_{\pm}\varepsilon\delta + \delta^2 = (2 + \alpha_{\pm}\delta)^2. \tag{6.20}$$

Then equation (6.18) reduces to

$$\omega_g^2/\omega_0^2 = [(2 + \alpha_{\pm}\delta)B_{11} + 1]^2. \tag{6.21}$$

It is equivalent to

$$\omega_g/\omega_0 = |(2 + \alpha_{\pm}\delta)B_{11} + 1|. \tag{6.22}$$

In the original notations, equation (6.22) becomes

$$\omega_g/\omega_0 = |\left\{1 + I_{22,0}/I_{11,0} \pm [(I_{22,0}/I_{11,0} - 1)^2 + I_{12,0}{}^2/I_{11,0}{}^2]^{1/2}\right\}B_{11}(E) + 1|. \quad (6.23)$$

The outcome of equation (6.23) is that there is a value of $B_{11}(E)$ (for each direction of the revolution) such that the generalized frequency ω_g vanishes. Consequently the radiation also vanishes. Here the explicit values $B_{11+}(E_{st})$ and $B_{11-}(E_{st})$ of $B_{11}(E)$, corresponding to such non-radiating (stationary) states (the subscript 'st' stands for 'stationary'), are

$$B_{11+}(E_{st}) = -1/\left\{1 + I_{22,0}/I_{11,0} + [(I_{22,0}/I_{11,0} - 1)^2 + I_{12,0}{}^2/I_{11,0}{}^2]^{1/2}\right\} \quad (6.24)$$

for $\alpha = \alpha_+$ and

$$B_{11-}(E_{st}) = -1/\left\{1 + I_{22,0}/I_{11,0} - [(I_{22,0}/I_{11,0} - 1)^2 + I_{12,0}{}^2/I_{11,0}{}^2]^{1/2}\right\} \quad (6.25)$$

for $\alpha = \alpha_-$.

Figure 6.1 presents the quantity B_{11+} (denoted in the plot for brevity as B_+) versus $I_{22,0}/I_{11,0}$ (denoted in the plot as C) and $I_{12,0}/I_{11,0}$ (denoted in the plot as D).

Figure 6.2 presents the quantity B_{11-} (denoted in the plot for brevity as B_-) versus $I_{22,0}/I_{11,0}$ (denoted in the plot as C) and $I_{12,0}/I_{11,0}$ (denoted in the plot as D).

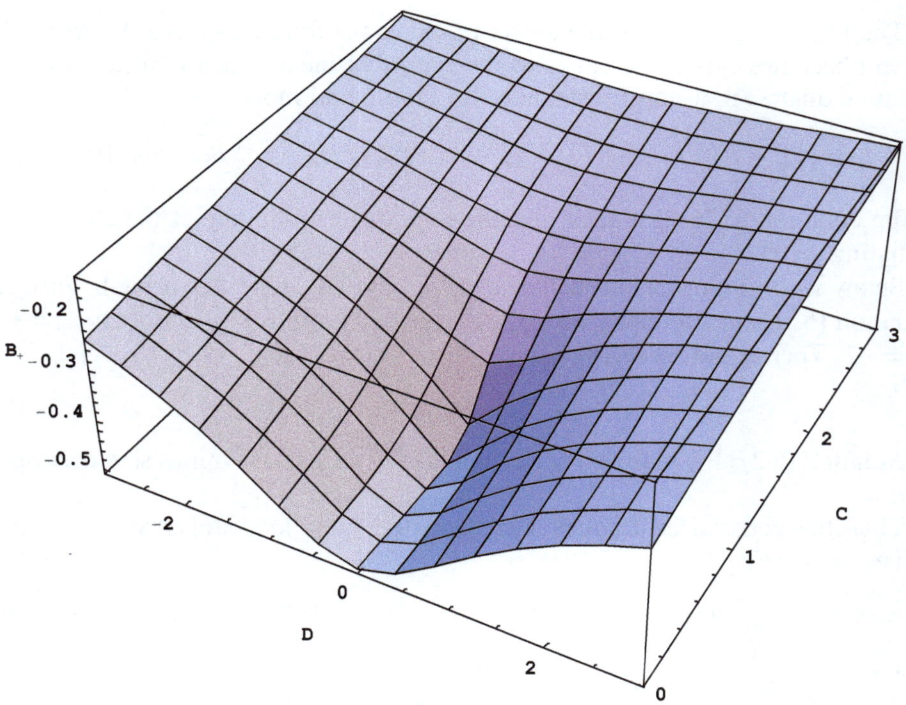

Figure 6.1. Plot of B_{11+} (denoted in the plot for brevity as B_+) from equation (6.24) versus $I_{22,0}/I_{11,0}$ (denoted in the plot as C) and $I_{12,0}/I_{11,0}$ (denoted in the plot as D). Reproduced from [11] with permission of MDPI.

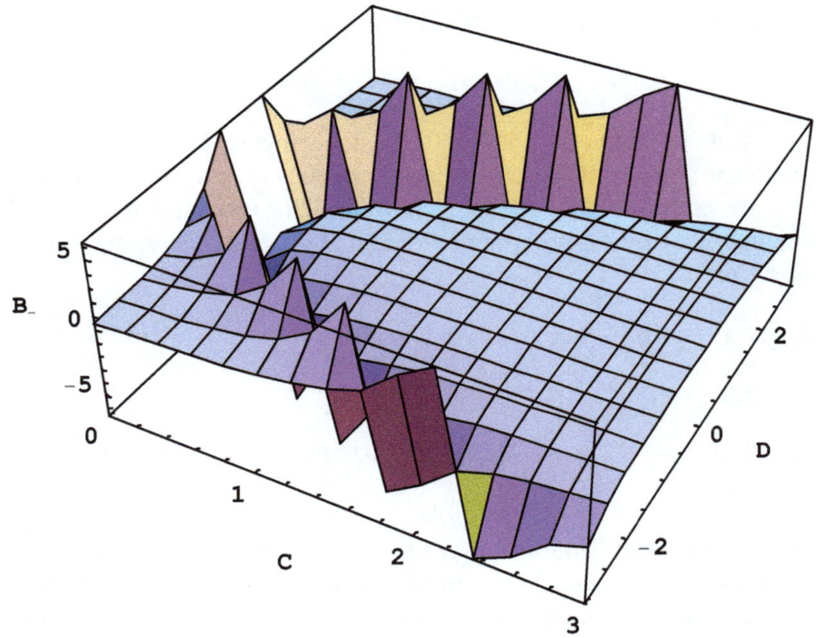

Figure 6.2. Plot of B_{11-} (denoted in the plot for brevity as B_{-}) from equation (6.25) versus $I_{22,0}/I_{11,0}$ (denoted in the plot as C) and $I_{12,0}/I_{11,0}$ (denoted in the plot as D). Reproduced from [11] with permission of MDPI.

The physical meaning of the above results is the following. As the energy of the system becomes closer and closer to the energy of the classical non-radiating state, the time dilates (in a non-Einsteinian way) more and more:

$$t' = |\left\{1 + I_{22,0}/I_{11,0} \pm [(I_{22,0}/I_{11,0} - 1)^2 + I_{12,0}{}^2/I_{11,0}{}^2]^{1/2}\right\} B_{11}(E) + 1|t. \tag{6.26}$$

From equation (6.26) it is seen that the time dilates infinitely at the classical non-radiating state.

Below is an important particular case, $I_{22,0} = I_{11,0}$ (circular orbits). Then from equation (6.16) one obtains $\gamma = 0$. Then from equation (6.17) we obtain $\alpha_{+} = 1$ and $\alpha_{-} = -1$. Then equation (6.18) reduces to

$$\omega_g/\omega_0 = |(2 \pm |I_{12,0}|/I_{11,0})B_{11}(E) + 1|. \tag{6.27}$$

In equation (6.27) the plus sign corresponds to $\alpha = 1$ and the minus sign corresponds to $\alpha = -1$.

Thus the generalized frequency ω_g and the radiation vanish at the following values of $B_{11}(E)$:

$$B_{11+}(E_{st}) = -1/(2 + |I_{12,0}|/I_{11,0}) \text{ for } \alpha = 1 \tag{6.28}$$

and

$$B_{11-}(E_{st}) = -1/(2 - |I_{12,0}|/I_{11,0}) \text{ for } \alpha = -1. \tag{6.29}$$

In the exceptional case where $|I_{12,0}|/I_{11,0} = 2$, equation (6.29) is not valid. Then equation (6.27) yields a trivial result: $\omega_g = \omega_0$.

The main result for the circular orbits are classical non-radiating states, corresponding to $B_{11}(E) = B_{11+}(E_{st})$ for $\alpha = 1$ or $B_{11}(E) = B_{11-}(E_{st})$ for $\alpha = -1$. For each direction of the revolution, there is only one value of $B_{11}(E)$, but this one value of $B_{11}(E)$ can entail many classical stationary states.

For example, the quantity B_{11} could have an oscillatory dependence on the energy, so that there would be an *infinite number* of energies of the classical stationary states E_{st}—in correspondence with the quantum solution.

An illustrative example:

$$B_{11+}(E) = -|\cos[\pi(E - C)/(E_{st,0} - C)]|/(2 + |I_{12,0}|/I_{11,0}). \tag{6.30}$$

Here $E_{st,0}$ is the energy of the ground state and E and $E_{st,0}$ are measured in units of $\hbar\omega_0$. The constant C in equation (6.30) is analogous to the Maslov index [12]. The Maslov index for spherically symmetric potentials is equal to 1/2 (see e.g. [13]). By using $C = 1/2$ we rewrite equation (6.30) as follows:

$$B_{11+}(E) = -|\cos[\pi(E - 1/2)/(E_{st,0} - 1/2)]|/(2 + |I_{12,0}|/I_{11,0}). \tag{6.31}$$

The consequence of equation (6.31) is that for $E = E_{st,n}$, where

$$E_{st,n} - 1/2 = (n + 1)(E_{st,0} - 1/2), \qquad n = 0, 1, 2, \ldots, \tag{6.32}$$

the quantity B_{11+} satisfies equation (6.18). Then the set of values $E_{st,n}$ from equation (6.32) is the set of the energies of the classical non-radiating stationary states:

$$E_{st,n} = (n + 1)E_{st,0} - n/2. \tag{6.33}$$

For $E_{st,0} = 3/2$ then the set of values $E_{st,n}$ from equation (6.33) is exactly the same as in the corresponding quantum results.

In closing, all classical spherically symmetric potentials have an additional vector integral of motion, for which the *quantal counterpart-operator does not exist* [14–16]. (This was utilized in [9]; Camarena and Oks [9] successfully applied the GHD to a modified Coulomb potential.) This presents possibilities that quantum mechanics does not have, as noted in the paper [8].

References

[1] Dirac P A M 1950 *Can. J. Math.* **2** 129

[2] Dirac P A M 1958 *Proc. R. Soc.* A **246** 326

[3] Dirac P A M 1964 *Lectures on Quantum Mechanics* (New York: Academic) reprinted by Dover 2001

[4] Sergi A 2004 *Phys. Rev.* E **69** 021109

[5] Sergi A 2005 *Phys. Rev.* E **72** 031104

[6] Sergi A 2005 *Phys. Rev.* E **72** 066125

[7] Sergi A 2006 *J. Chem. Phys.* **124** 024110

[8] Oks E and Uzer T 2002 *J. Phys. B: At. Mol. Opt. Phys.* **35** 165

[9] Camarena A and Oks E 2010 *Int. Rev. At. Mol. Phys.* **1** 143

[10] Oks E 2015 *Breaking Paradigms in Atomic and Molecular Physics* (Singapore: World Scientific) ch 4
[11] Oks E 2020 *Symmetry* **12** 1130
[12] Maslov V P and Fedoriuk M V 1981 *Semi-Classical Approximation in Quantum Mechanics* (Berlin: Springer)
[13] Landau L D and Lifshitz E M 1965 *Quantum Mechanics* (Oxford: Pergamon)
[14] Mukunda N 1967 *Phys. Rev. (Second Series)* **155** 1383
[15] Bacry H, Ruegg H and Souriau J M 1966 *Commun. Math. Phys.* **3** 323
[16] Fradkin D M 1967 *Progr. Theor. Phys.* **37** 798

Chapter 7

Selected applications for spectroscopic diagnostics of plasmas

7.1 Quasienergy states as a tool for diagnosing quasimonochromatic electric fields in plasmas

The development of methods for measuring oscillatory electric fields (OEFs) $\vec{E}_M(t) = \vec{E}_{0M} \cos \omega_M t$ in plasmas, for instance, microwave fields, is of great practical importance. First, these are fields used for producing microwave discharges. Second, the presence of fields $\vec{E}_M(t)$ in plasmas is typical for many experiments where interaction of a strong microwave radiation with the plasma is studied. Third, a microwave field $\vec{E}_M(t)$ is widely used for heating tokamak plasmas. In the case where the laser spectroscopy is used for measuring the OEF $\vec{E}_M(t) = \vec{E}_{0M} \cos \omega_M t$, atoms in a plasma interact with two electric fields: a laser field $\vec{E}_L(t) = \vec{E}_{0L} \cos \omega_L t$ and the OEF $\vec{E}_M(t)$ (we assume $\omega_M \ll \omega_L$). In this chapter we follow our conference presentation [1] to describe the underlying physical principles that can serve as a basis for laser-aided diagnostics of OEFs in a plasma.

Let the frequency ω_L be resonant to a transition between upper (a) and lower (b) atomic levels: $\omega_L \approx \omega_{ab}$. In this situation the OEF $\vec{E}_M(t)$ can modify states belonging separately to the levels a and/or b. However, the OEF $\vec{E}_M(t)$ alone does not induce the transitions between levels a and b. Let the frequency ω_M be much greater than the impact width, i.e. the full width at half maximum (FWHM) γ_{impact}, and the Doppler width γ_D of the spectral line $a \leftrightarrow b$. In this case the principles for measuring the OEF $\vec{E}_M(t)$ can be based on the use of the multi-quantum resonant transitions between levels a and b. Such transitions may occur with the participation of one quantum $\hbar\omega_L$ of the laser field and k quanta $\hbar\omega_M$ of the OEF $\vec{E}_M(t)$: $\omega_L \approx \omega_{ab} + k\omega_M$, $k = 0, \pm1, \pm2, \ldots$..

First we discuss the absorption spectra of atoms without permanent dipole moments. The interaction of an atom with the OEF $\vec{E}_M(t)$ can be characterized by the parameter

$$R = |d_{\alpha'\alpha''}E_{0M}/[2\hbar(\omega_{\alpha'\alpha''} \pm \omega_M)]|, \tag{7.1}$$

where $d_{\alpha'\alpha''}$ is the matrix element of the dipole moment between the closely spaced levels α', α'', and $\omega_{\alpha'\alpha''}$ is the separation between these levels. If $R \ll 1$ (the case of a weak field $\vec{E}_M(t)$), the field $\vec{E}_M(t) = \vec{E}_{0M} \cos \omega_M t$ can be measured using the fact that the excitation rate for the two-quantum process, involving one quantum $\hbar\omega_L$ of the laser field $\vec{E}_L(t) = \vec{E}_{0L} \cos \omega_L t$ and one quantum $\hbar\omega_M$ of the field $\vec{E}_M(t) = \vec{E}_{0M} \cos \omega_M t$, is proportional to E_{0M}^2. Let us consider, for instance, the excitation scheme shown in figure 7.1, where $\omega_{23} \ll \omega_{21}$. We assume that in the electric dipole approximation, the single-photon transitions $1 \leftrightarrow 3$ and $2 \leftrightarrow 3$ are allowed, whereas the single-photon transition $1 \leftrightarrow 2$ is forbidden.

Using the quantum mechanical perturbation theory (we assume that $R \ll 1$ for levels 2 and 3), one can obtain the excitation rates for the two quantum transitions shown in figure 7.1:

$$\begin{aligned} W_{1\to2}^{(+)} &= \left\{E_{0L}^2 E_{0M}^2 d_{13}^2 d_{32}^2/[8\hbar^2(\omega_{32} + \omega_M)^2]\right\} L(\omega_{21} - \omega_L - \omega_M, \gamma), \\ W_{1\to2}^{(-)} &= \left\{E_{0L}^2 E_{0M}^2 d_{13}^2 d_{32}^2/[8\hbar^2(\omega_{32} - \omega_M)^2]\right\} L(\omega_{21} - \omega_L + \omega_M, \gamma). \end{aligned} \tag{7.2}$$

In equation (7.2) $L(\omega, \gamma) \equiv (\gamma/2)/(\omega^2 + \gamma^2/4)$, γ is the width (FWHM) of the line. The value of $W_{1\to2}^{(+)}$ corresponds to the process shown in figure 7.1(a), whereas the value of $W_{1\to2}^{(-)}$ corresponds to the process shown in figure 7.1(b). The excitation rate for the direct single-photon transition $1 \to 3$ induced by the laser field is as follows:

$$W_{1\to3} = \left[d_{13}^2 E_{0L}^2/(2\hbar^2)\right] L(\omega_{21} - \omega_L, \gamma). \tag{7.3}$$

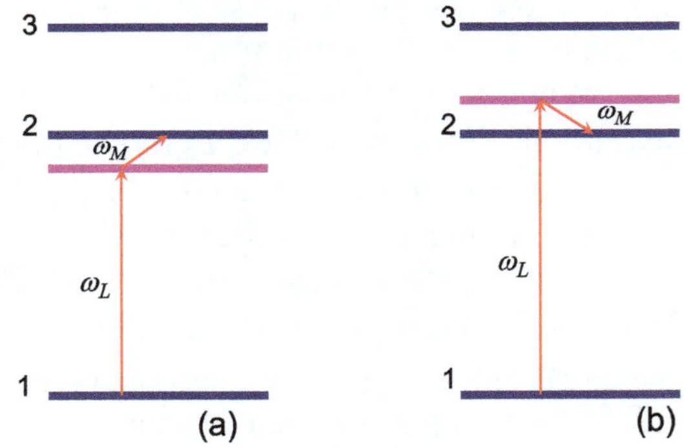

Figure 7.1. Partial scheme of energy levels of an atom. Reproduced with permission from [1]. (a) The excitation of the transition $1 \to 2$ due to the absorption of a laser photon $\hbar\omega_L$ and of an OEF quantum $\hbar\omega_M$. (b) The excitation of the transition $1 \to 2$ due to the absorption of a laser photon $\hbar\omega_L$ and the emission of an OEF quantum $\hbar\omega_M$.

Figure 7.2 shows the absorption spectrum consisting of the lines $W^{(\pm)}_{1\to 2}$ and $W_{1\to 3}$ for the transitions 1 → (2, 3). One can see that both the ratios $W^{(+)}_{1\to 2}/W_{1\to 3}$ and the ratio $W^{(-)}_{1\to 2}/W_{1\to 3}$ are proportional to E^2_{0M}. Therefore, the ratios $W^{(+)}_{1\to 2}/W_{1\to 3}$ and $W^{(-)}_{1\to 2}/W_{1\to 3}$ can be used to measure E_{0M} by scanning the laser frequency around the frequencies ω_{21} and ω_{31} and observing the fluorescence signals at transitions originated from levels 2 and 3.

In the case where the parameter R in equation (7.1) is of the order or greater than unity, the formalism of quasienergy states (QES) is more appropriate for treating the interaction of the atom with the OEF $\vec{E}_{\mathrm{M}}(t)$. According to the Floquet theory (see e.g. [2]), for the atom interacting with the OEF $\vec{E}_{\mathrm{M}}(t) = \vec{E}_{0\mathrm{M}} \cos \omega_{\mathrm{M}} t$, among the wave functions of the atom, there exists a basis of the wave functions of QES

$$\Phi_n(\vec{r}, t) = \exp(-\mathrm{i}\varepsilon_n t)\sum_k \exp(-\mathrm{i}k\omega_{\mathrm{M}} t)\phi_{nk}(\vec{r}), \tag{7.4}$$

where ε_n is the quasienergy of level n and the functions $\phi_{nk}(\vec{r})$ are time independent. We note that the term 'quasienergy states' was introduced in papers [3, 4]. The quasienergies ε_n and the functions $\phi_{nk}(\vec{r})$ can be obtained either numerically, or analytically by solving the Schrödinger equation. Generally, they depend on the parameters of the field $\vec{E}_{\mathrm{M}}(t) = \vec{E}_{0\mathrm{M}} \cos \omega_{\mathrm{M}} t$. Taking into account equation (7.4), one can see that the excitation of the transition $b \to a$ can occur when the following resonance condition is satisfied:

$$\omega_{\mathrm{L}} \approx (\varepsilon_a - \varepsilon_b)/\hbar + k\omega_{\mathrm{M}}, \quad k = 0, \pm 1, \pm 2, \ldots,$$

where ε_a and ε_b are the quasienergies of levels a and b, respectively. Using equation (7.4) we find that for determining the excitation rate for the transition $b \to a$ we can use equation (7.3) in which the dipole matrix element d_{13} should be replaced by the effective dipole matrix element $D^{(k)}_{ba} = \langle \varphi_b | d | \phi_{ak} \rangle$, where φ_b is the wave function of

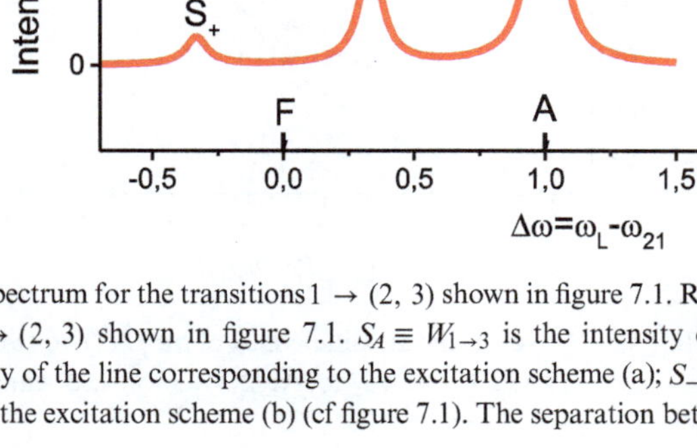

Figure 7.2. Absorption spectrum for the transitions 1 → (2, 3) shown in figure 7.1. Reproduced with permission from [1]. Transitions 1 → (2, 3) shown in figure 7.1. $S_A \equiv W_{1\to 3}$ is the intensity of the allowed line 1 → 3; $S_+ \equiv W^{(+)}_{1\to 2}$ is the intensity of the line corresponding to the excitation scheme (a); $S_- \equiv W^{(-)}_{1\to 2}$ is the intensity of the line corresponding to the excitation scheme (b) (cf figure 7.1). The separation between the lines S_- and S_+ is equal to $2\omega_M$.

the lower level b. (We assume here that the strong OEF $\vec{E}_M(t)$ perturbs only the upper state a.) As a result, the excitation rate for the transition $b \to a$ can be expressed as

$$W_{b\to a} = [|<\varphi_b|d|\phi_{ak}>|^2 E_{0L}^2/(2\hbar^2)]L((\varepsilon_a - \varepsilon_b)/\hbar - \omega_L + k\omega_M, \gamma).$$

Let us assume that levels b and a coincide, respectively, with levels 1 and 2 shown in figure 7.1. In the case of a weak OEF $\vec{E}_M(t)$, according to equation (7.3), the ratio $W^{(+)}_{1\to 2}/W^{(-)}_{1\to 2}$ does not depend on E_{0M}. However, for the strong OEF $\vec{E}_M(t)$, this ratio is a function of the parameters of the OEF. Figure 7.3 shows, as an example, the ratio $W^{(+)}_{1\to 2}/W^{(-)}_{1\to 2}$ versus the amplitude of the OEF $\vec{E}_M(t)$, calculated for two two-quantum transitions in helium atoms: the transitions $2^1P \to 4^1F$ and $2^3P \to 4^3F$. For helium atoms, levels $2^{2S+1}P$, $4^{2S+1}F$, and $4^{2S+1}D$ ($S = 0;\ 1$) play the role of levels 1, 2, and 3 shown in figure 7.1. Using the dependence on the parameters of the OEF, one can measure these parameters in a plasma by observing the laser-induced fluorescence signal at the transition from level 2 to some lower level q. The advantage of this method is that both the fluorescent lines originate from the same level 2—therefore, a possible collisional transfer of the population from level 2 to nearby levels does not interfere with the measurements of the OEF.

Now we discuss the absorption spectra of atoms possessing permanent dipole moments: ($d_{aa} \equiv <\varphi_a|d|\varphi_a>\neq 0$). Let us consider laser-induced transitions in such atoms possessing permanent dipole moments. Hydrogen atoms are one example of such atoms. In the case where $d_{aa} \neq 0$, the wave function of a QES of the atom, which is in the state a and interacts with the OEF $\vec{E}_M(t) = \vec{E}_{0M} \cos \omega_M t$, has the form [5]

$$\Psi_a(\vec{r}, t) = \exp[-id_{aa} E_{0M} \sin \omega_M t/(\hbar\omega_M)]\varphi_a(\vec{r}). \tag{7.5}$$

Let the frequency ω_L of the laser radiation be close to the frequency of the transition $b \leftrightarrow a$. Performing the Fourier series expansion in equation (7.5) and taking into

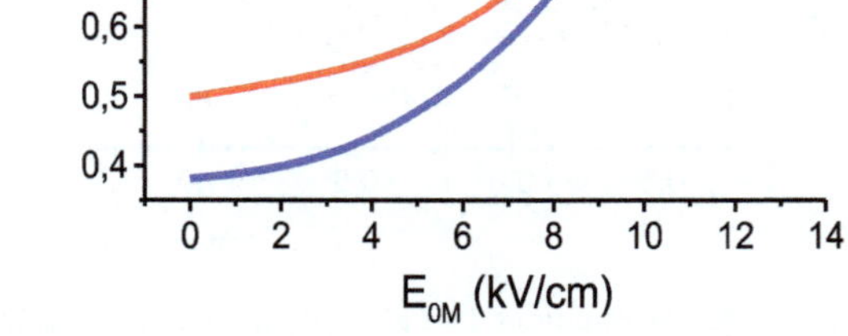

Figure 7.3. The ratio $Y = W^{(+)}_{1\to 2}/W^{(-)}_{1\to 2}$ versus the amplitude E_{0M} of the OEF for two two-quantum transitions in helium atoms: $2^1P \to 4^1F$ (curve 1) and $2^3P \to 4^3F$ (curve 2). Reproduced with permission from [1]. The OEF frequency in this example ω_M was 2.4×10^{11} s^{-1}.

account the formula $\exp(-ix \sin \omega_M t) = \sum_{n=-\infty}^{+\infty} J_n(x)\exp(-in\omega_M t)$, where $J_n(x)$ are the Bessel functions, we find that the effective matrix element of the dipole moment between levels a and b has the form $D_{ab}^{(k)} = J_k(\Delta\beta_{ab})d_{ab}$, where $\Delta\beta_{ab} \equiv (d_{aa} - d_{bb})E_{0M}/(\hbar\omega_M)$. Using the above expression for $D_{ab}^{(k)}$, we obtain the increase of the population ΔN_a of the upper level a

$$\begin{aligned} \Delta N_a &\equiv N_a - N_{a0} = 2^{-1}G_k(N_{b0} - N_{a0}) \\ &\quad \Big/\left[1 + (\omega_{ab} + k\omega_M - \omega_L - \sigma_k)^2\tau_{12}^2 + G_k\right], \\ G_k &\equiv 4W_{ab}^2 J_k^2(\Delta\beta_{ab})\tau_{12}\Gamma^{-1}, \quad \sigma_k = 2\omega_M^{-1}W_{ab}^2 g_k(\Delta\beta_{ba}), \\ g_k(\nu) &= \sum_{r=1}^{\infty}\left[J_{k-r}^2(\nu) - J_{k+r}^2(\nu)\right]\Big/ r, \quad W_{ab} \equiv d_{ab}E_{0L}/(2\hbar). \end{aligned} \tag{7.6}$$

Here τ_{12} and Γ^{-1} are transverse and longitudinal relaxation times, respectively. The quantity τ_{12} is inversely proportional to the impact width of the spectral line $a \to b$; the quantity Γ controls the relaxation rate of levels a and b. From equation (7.6) it follows that the spectrum of the absorption of the laser field by atoms consists of satellites at the frequencies

$$\omega_L = \omega_{ab} + k\omega_M - \sigma_k, \quad k = 0, \ \pm 1, \ \pm 2, \ \ldots, \tag{7.7}$$

the half-width of the kth satellite being $\Delta\omega_{k,1/2} = 2(1 + G_k)^{1/2}/\tau_{12}$. We point out that the term σ_k in the denominator of the expression for ΔN_a (see equation (7.6)) is related to the dynamic Stark shift of the quasienergy levels. It depends on both the laser field and the OEF $\vec{E}_M(t)$. We note that the expression for σ_k was first obtained in papers [6, 7]. Figure 7.4 shows, as an example, the absorption spectrum of the laser radiation by atoms possessing the permanent dipole moments in the states a and b. It consists of the unshifted line S_0 and of the satellite lines $S_{\pm k}$ ($k = 1, 2, \ldots$).

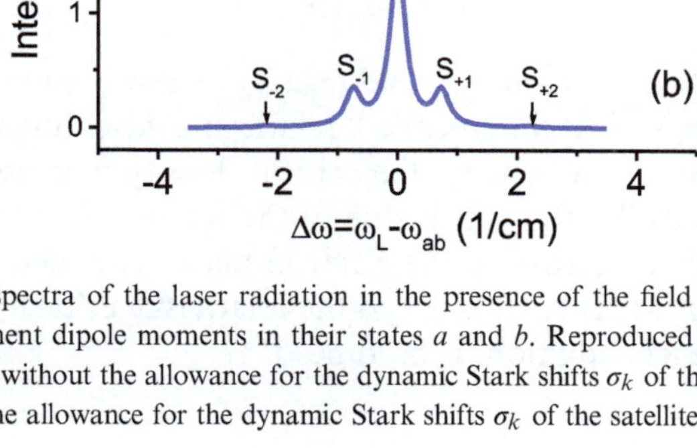

Figure 7.4. Absorption spectra of the laser radiation in the presence of the field $E_M(t) = E_{0M}\cos\omega_M t$ by atoms possessing permanent dipole moments in their states a and b. Reproduced with permission from [1]. Profile (a) was calculated without the allowance for the dynamic Stark shifts σ_k of the satellites, whereas profile (b) was calculated with the allowance for the dynamic Stark shifts σ_k of the satellites.

The intensity of these lines, according to equation (7.6), is the function of the amplitude $E_{0\mathrm{M}}$ of the OEF.

For figure 7.4 the matrix elements of the dipole moment were assumed to be as follows: $d_{ab} = 2.15ea_0$, $d_{aa} - d_{bb} = 3.5ea_0$, where a_0 is the Bohr radius. Profiles shown in figure 7.4 were calculated at $E_{0\mathrm{L}} = 15$ kV cm^{-1}, $E_{0M} = 7$ kV cm^{-1}, and $\omega_\mathrm{M} = 2.4 \times 10^{11}$ s^{-1}. For the profile (a), the separation between the satellites S_{+k} and S_{-k} is equal to $2k\omega_\mathrm{M}$. One can see from figure 7.4 that the allowance for the dynamic Stark shifts σ_k can be important for calculations of the absorption spectra of atoms.

Equation (7.6) can be used for measuring the amplitude of the microwave field $E_{0\mathrm{M}}$ in plasmas. For this purpose one should record the wavelength-integrated intensity of the laser-induced fluorescence $I_\mathrm{f} \propto \Delta N_2$ during the transition from the upper level a to one of the lower levels versus the laser field intensity $I_\mathrm{L} \propto E_{0\mathrm{L}}^2$: $I_\mathrm{f}^{-1} \propto (1 + I_{k,\mathrm{satur}} I_\mathrm{L}^{-1})$, $I_{k,\mathrm{satur}} I_\mathrm{L}^{-1} \equiv G_k^{-1}(E_{0\mathrm{M}})$. Here k is the number of microwave quanta involved in the resonance (see equation (7.7)). The amplitude $E_{0\mathrm{M}}$ can be measured in two ways. In the first, one could measure the ratio of slopes of experimental dependences $I_\mathrm{f}^{-1}(I_\mathrm{L}^{-1})$ at two different values of $k = 0$ ($k = k'$ and $k = k''$), and use the dependence of the ratio $G_{k'}/G_{k''}$ on $E_{0\mathrm{M}}$. In the second, one could use the dependence of the dynamic Stark shift σ_k on the parameters of the OEF $\vec{E}_\mathrm{M}(t)$. By scanning the laser frequency around the frequency ω_{ab} and by recording peaks of the laser-induced fluorescence intensity, one could tune to multi-quantum resonances (7.7) corresponding to two different indices $k = k_1$ and $k = k_2$. The laser frequency should be kept constant. In this situation, one would have

$$\omega_{L,1} = \omega_{ab} + k_1\omega_\mathrm{M} - \sigma_{k_1}, \qquad \omega_{L,2} = \omega_{ab} + k_2\omega_\mathrm{M} - \sigma_{k_2}. \tag{7.8}$$

Since ω_M is known, from equation (7.8) one would determine the dynamic Stark shifts σ_{k_1} and σ_{k_2}. Then, using the fact that the ratio $\sigma_{k_2}/\sigma_{k_1}$ depends on $E_{0\mathrm{M}}$ but does not depend on $E_{0\mathrm{L}}$, one could finally find the OEF amplitude $E_{0\mathrm{M}}$.

7.2 Measuring the laser field and the opacity of spectral lines in plasmas

In experimental studies of laser–plasma interactions, the laser radiation cannot penetrate plasma regions (called 'overdense' plasma) of the electron density $N_\mathrm{e} > N_\mathrm{c}$, where N_c is the critical density defined by the equation

$$\omega = \omega_{pe}(N_c). \tag{7.9}$$

Here ω is the laser/maser frequency and $\omega_\mathrm{pe}(N_\mathrm{c})$ is the plasma electron frequency: $\omega_\mathrm{pe}(N_\mathrm{c}) = (4\pi e^2 N_\mathrm{c}/m_\mathrm{e})^{1/2}$. At the so-called 'relativistic' laser intensities—typically the intensities exceeding 10^{18} W cm^{-2}—the critical density increases (despite the laser frequency being fixed) [8–10]. This is due to the relativistic effects in the plasma—such as the relativistic increase of the electron mass. The increased critical density depends on the laser intensity. It is called the relativistic critical density N_cr. For the linearly polarized laser radiation, it becomes [11]

$$N_{\mathrm{cr}} = \frac{(\pi a/4) m_{\mathrm{e}}\omega^2}{4\pi e^2}, \quad a = \lambda\ (\mu m)\left[\frac{I\ (\mathrm{W\ cm^{-2}})}{1.37 \times 10^{18}}\right]^{1/2}. \tag{7.10}$$

Below we utilize the term 'critical density' in the broader sense—including the relativistic critical density.

However, the laser radiation can exist inside plasma regions where the electron density is below the critical density ('underdense' plasma), as well as at the surface of the critical density. The surface of the critical density could exhibit a rich physics. Namely, the incident laser radiation can be converted into transverse electromagnetic waves of significantly higher amplitudes than the incident radiation—due to various nonlinear processes.

The laser fields (or more generally, the field of the resulting transverse electromagnetic wave) in laser-produced plasmas were measured in experiments using satellites of dipole-forbidden spectral lines of helium-like ions—see e.g. [12, 13] and references therein. The satellites of hydrogenic ions have not been used for this purpose yet (to the best of our knowledge), despite the underlying theory being well-developed—see e.g. [5, 14] and [15, 16]. One of the reasons is the following. Typically, the laser field would be determined from the experimental ratio of the satellite intensity to the intensity of the main line (the line at the 'unperturbed' wavelength or frequency) by comparing it with the corresponding theoretical ratio. However, in the very dense plasmas characteristic of laser–plasma interactions (in particular for relativistic laser–plasma interactions), intense hydrogenic lines, such as the Ly-alpha and Ly-beta lines, could be optically thick. In this situation the experimental peak intensity of the main line would be affected by the opacity (while the peak intensity of the satellites would not be affected), so that the existing theory cannot be used for deducing the laser field from the experimental ratio of the satellite intensity to the intensity of the main line.

In the present chapter, we propose a method that is appropriate for the above situation—the method that allows measuring both the laser field and the opacity from experimental spectrum of a hydrogenic line exhibiting satellites. We obtain the necessary theoretical results analytically and show how to use them for this purpose.

Under a linearly polarized electric field $\mathbf{E}_0\cos\omega t$, a Stark component of a hydrogenic spectral line splits in satellites separated by $p\omega$ ($p = \pm1, \pm2, \pm3, \ldots$) from the unperturbed frequency ω_0 of the spectral line, as shown by Blochinzew [5]. Blochinzew's result was later extended to profiles of multicomponent hydrogenic spectral lines in [14]. In the 'reduced frequency' scale, the profile is as follows (presented also in [15, section 3.1]):

$$S(\Delta\omega/\omega) = \sum_{p=-\infty}^{+\infty} I(p,\ \varepsilon)\ \delta((\Delta\omega/\omega) - p),$$
$$I(p,\ \varepsilon) = [f_0\ \delta_{p0} + 2\sum_{k=1}^{\mathrm{kmax}} f_k\ J_p^2(X_k\varepsilon)]/(f_0 + 2\,\Sigma f_k), \quad X_k = nq - n_0q_0. \tag{7.11}$$

Here n, q and n_0, q_0 are the principal and electric quantum numbers of the upper and lower energy levels, respectively, involved in the radiative transition, f_0 is the total intensity of all central Stark components, f_k is the intensity of the lateral Stark component with the number $k = 1, 2, \ldots, k_{max}$, $J_p(z)$ are the Bessel functions, and ε is the scaled dimensionless amplitude of the field:

$$\varepsilon = 3\hbar E_0/(2Z_r m_e e\omega). \tag{7.12}$$

In equation (7.11) Z_r is the nuclear charge of the radiating atom or ion, and m_e and e are the electron mass and charge, respectively. A practical formula for the scaled dimensionless amplitude of the field is the following:

$$\varepsilon = 1.204 \times 10^7\, E_0(V\ \mathrm{cm}^{-1})/[Z_r\ \omega(s^{-1})]. \tag{7.13}$$

(We recall that the electric quantum numbers q and q_0 are the differences of the corresponding parabolic quantum numbers, for example, $q = n_1 - n_2$.)

In the current chapter, we focus on the situation where $\varepsilon < 1$. In this situation, the intensities of the first satellite ($|p| = 1$) and of the second satellite ($|p| = 2$) are significantly smaller than the intensity of the main line, the latter being the zeroth satellite ($p = 0$). Therefore, even if the optical depth τ_0 at the main line were to be significant, the satellites ($|p| > 0$) typically would be optically thin.

According to equation (7.10) the ratio of the intensity of the second satellite to the intensity of the first satellite has the form

$$R_{21}(\varepsilon) = I(2,\ \varepsilon)/I(1,\ \varepsilon). \tag{7.14}$$

This ratio depends only on the scaled dimensionless amplitude ε of the electric field. Therefore, from the experimental value of the ratio R_{21}, one can determine the value of ε and then (by using equation (7.13)) the laser amplitude E_0.

As for the ratio of the intensity of the first satellite to the intensity of the main line, at the zero optical depth of the main line, it would be

$$R_{10}(\tau_0 = 0,\ \varepsilon) = I(1,\ \varepsilon)/\ I(0,\ \varepsilon). \tag{7.15}$$

Below, we present graphically the dependence of the ratios R_{21} and R_{10} on the scaled dimensionless laser field ε at $\tau_0 = 0$.

Figure 7.5 presents the dependence of the ratio $R_{21}(\varepsilon)$ of the intensity of the second satellite to the intensity of the first satellite, for the Lyman-beta line for the observation perpendicular to the laser field.

Figure 7.6 presents the dependence of the ratio $R_{10}(\varepsilon)$ of the intensity of the first satellite to the intensity of the main line, for the Lyman-beta line for the observation perpendicular to the laser field.

Figure 7.7 presents the dependence of the ratio $R_{21}(\varepsilon)$ of the intensity of the second satellite to the intensity of the first satellite, for the Lyman-beta line for the observation parallel to the laser field.

Figure 7.8 presents the dependence of the ratio $R_{10}(\varepsilon)$ of the intensity of the first satellite to the intensity of the main line, for the Lyman-beta line for the observation parallel to the laser field.

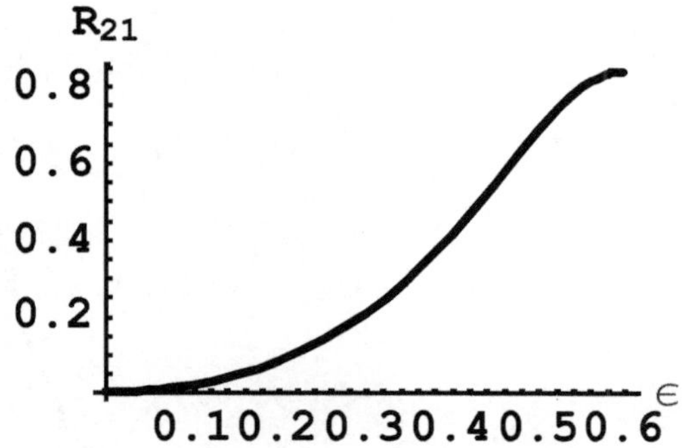

Figure 7.5. Ratio R_{21} of the intensity of the second satellite to the intensity of the first satellite versus the scaled dimensionless laser field ε, for the Lyman-beta line for the observation perpendicular to the laser field.

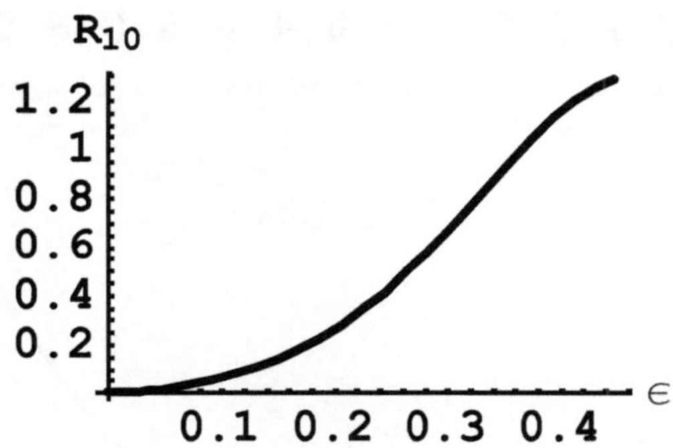

Figure 7.6. Ratio R_{10} of the intensity of the first satellite to the intensity of the main line versus the scaled dimensionless laser field ε, for the Lyman-beta line for the observation perpendicular to the laser field.

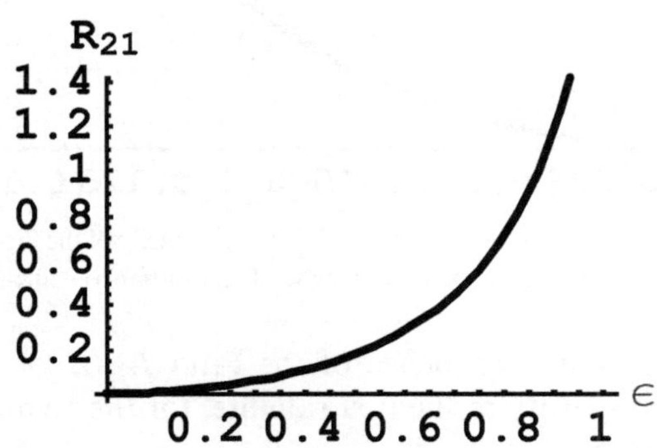

Figure 7.7. The same as in figure 7.5, but for the observation parallel to the laser field.

Figure 7.9 presents the dependence of the ratio $R_{21}(\varepsilon)$ of the intensity of the second satellite to the intensity of the first satellite, for the Lyman-delta line for the observation perpendicular to the laser field.

Figure 7.10 presents the dependence of the ratio $R_{10}(\varepsilon)$ of the intensity of the first satellite to the intensity of the main line, for the Lyman-delta line for the observation perpendicular to the laser field.

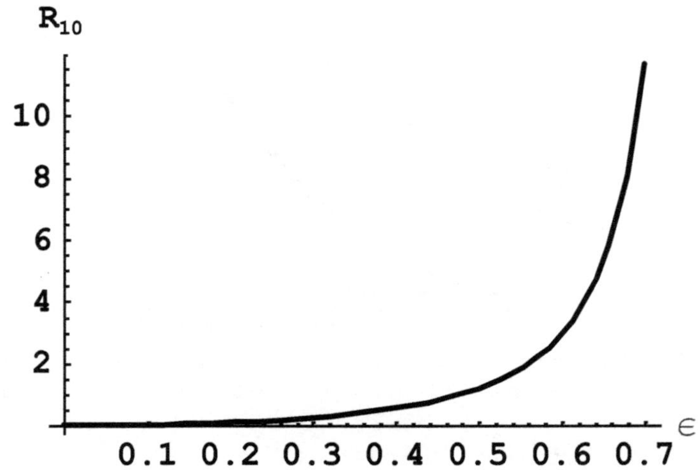

Figure 7.8. The same as in figure 7.6, but for the observation parallel to the laser field.

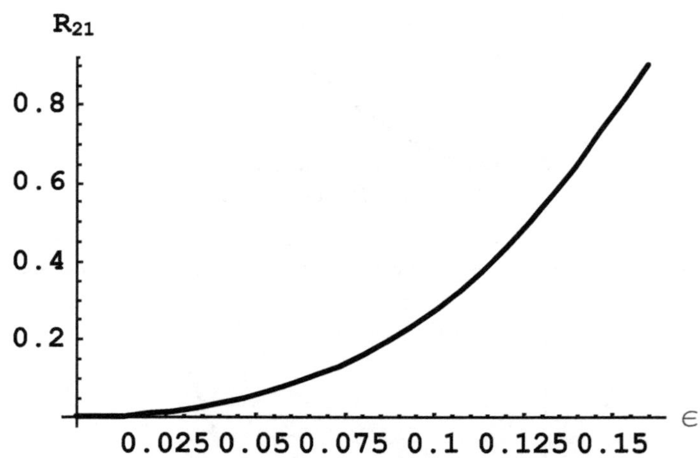

Figure 7.9. Ratio R_{21} of the intensity of the second satellite to the intensity of the first satellite versus the scaled dimensionless laser field ε, for the Lyman-delta line for the observation perpendicular to the laser field.

Figure 7.11 presents the dependence of the ratio $R_{21}(\varepsilon)$ of the intensity of the second satellite to the intensity of the first satellite, for the Lyman-delta line for the observation parallel to the laser field.

Figure 7.12 presents the dependence of the ratio $R_{10}(\varepsilon)$ of the intensity of the first satellite to the intensity of the main line, for the Lyman-delta line for the observation parallel to the laser field.

Figure 7.13 presents the dependence of the ratio $R_{21}(\varepsilon)$ of the intensity of the second satellite to the intensity of the first satellite, for the Balmer-beta line for the observation perpendicular to the laser field.

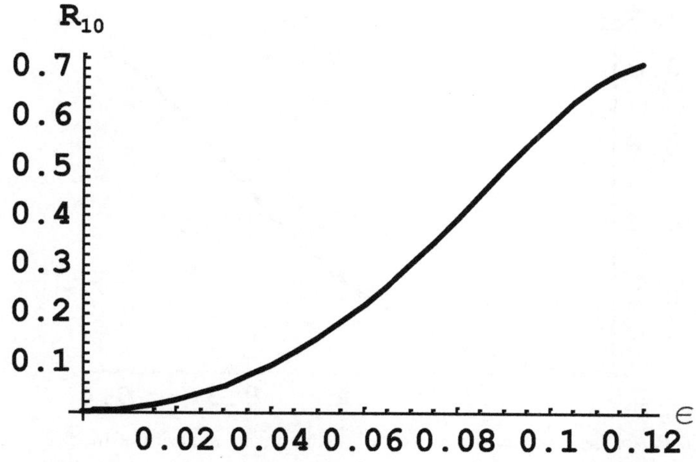

Figure 7.10. Ratio R_{10} of the intensity of the first satellite to the intensity of the main line versus the scaled dimensionless laser field ε, for the Lyman-delta line for the observation perpendicular to the laser field.

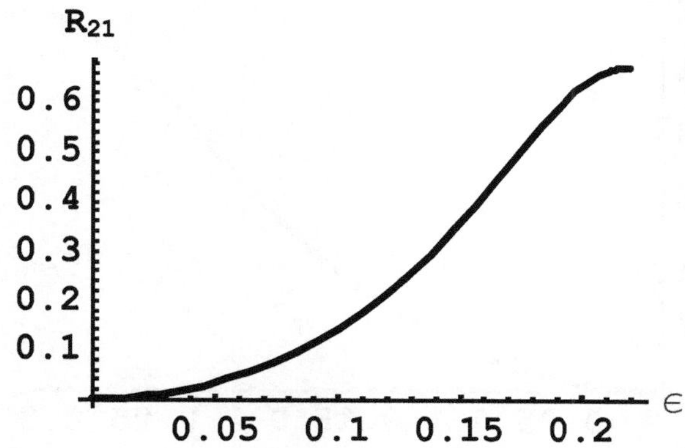

Figure 7.11. The same as in figure 7.9, but for the observation parallel to the laser field.

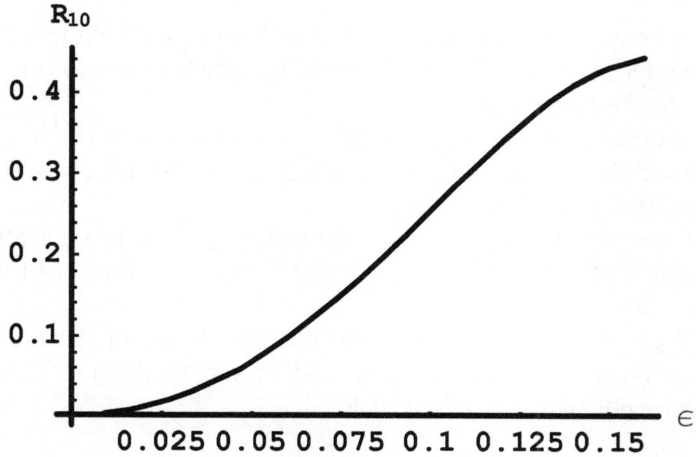

Figure 7.12. The same as in figure 7.10, but for the observation parallel to the laser field.

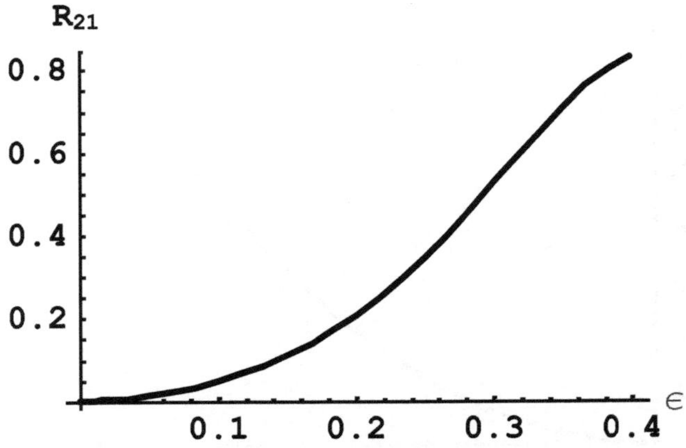

Figure 7.13. Ratio R_{21} of the intensity of the second satellite to the intensity of the first satellite versus the scaled dimensionless laser field ε, for the Balmer-beta line for the observation perpendicular to the laser field.

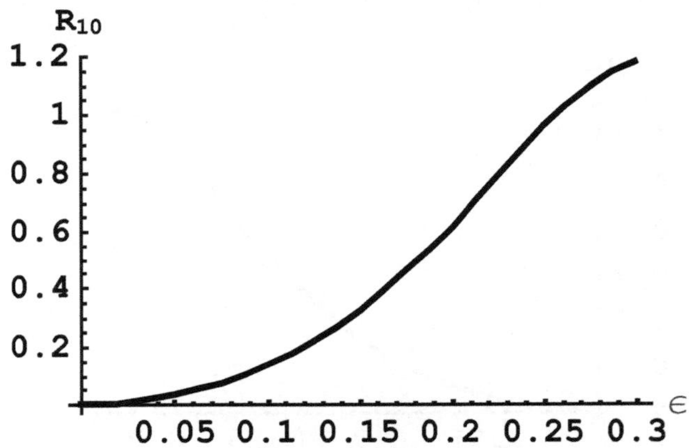

Figure 7.14. Ratio R_{10} of the intensity of the first satellite to the intensity of the main line versus the scaled dimensionless laser field ε, for the Balmer-beta line for the observation perpendicular to the laser field.

Figure 7.14 presents the dependence of the ratio $R_{10}(\varepsilon)$ of the intensity of the first satellite to the intensity of the main line, for the Balmer-beta line for the observation perpendicular to the laser field.

Figure 7.15 presents the dependence of the ratio $R_{21}(\varepsilon)$ of the intensity of the second satellite to the intensity of the first satellite, for the Balmer-beta line for the observation parallel to the laser field.

Figure 7.16 presents the dependence of the ratio $R_{10}(\varepsilon)$ of the intensity of the first satellite to the intensity of the main line, for the Balmer-beta line for the observation parallel to the laser field.

Figure 7.17 presents the dependence of the ratio $R_{21}(\varepsilon)$ of the intensity of the second satellite to the intensity of the first satellite, for the Balmer-delta line for the observation perpendicular to the laser field.

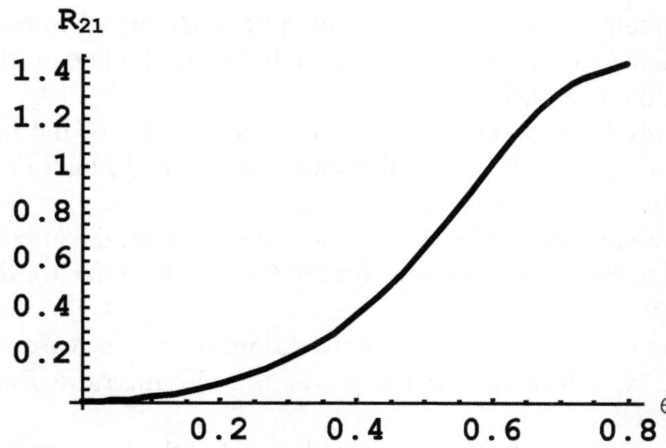

Figure 7.15. The same as in figure 7.13, but for the observation parallel to the laser field.

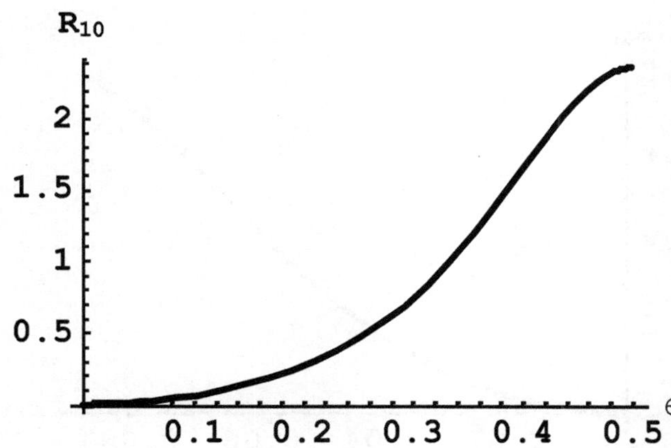

Figure 7.16. The same as in figure 7.14, but for the observation parallel to the laser field.

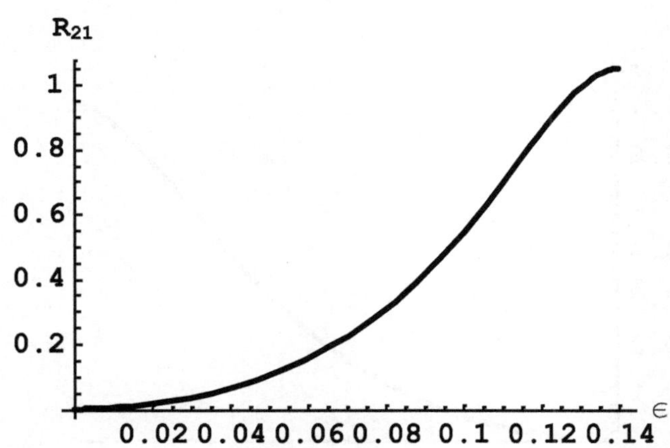

Figure 7.17. Ratio R_{21} of the intensity of the second satellite to the intensity of the first satellite versus the scaled dimensionless laser field ε, for the Balmer-delta line for the observation perpendicular to the laser field.

Figure 7.18 presents the dependence of the ratio $R_{10}(\varepsilon)$ of the intensity of the first satellite to the intensity of the main line, for the Balmer-delta line for the observation perpendicular to the laser field.

Figure 7.19 presents the dependence of the ratio $R_{21}(\varepsilon)$ of the intensity of the second satellite to the intensity of the first satellite, for the Balmer-delta line for the observation parallel to the laser field.

Figure 7.20 presents the dependence of the ratio $R_{10}(\varepsilon)$ of the intensity of the first satellite to the intensity of the main line, for the Balmer-delta line for the observation parallel to the laser field.

Now, we present the corresponding analytical results for the cases of the nonzero optical depth. At $\tau_0 > 0$ (and in particular at $\tau_0 > 1$), the ratio $R_{10}(\tau_0)$ has to be calculated as follows.

In equation (7.11) the profiles of each satellite and of the main line are represented by the delta-function. In reality, these profiles are influenced by various broadening

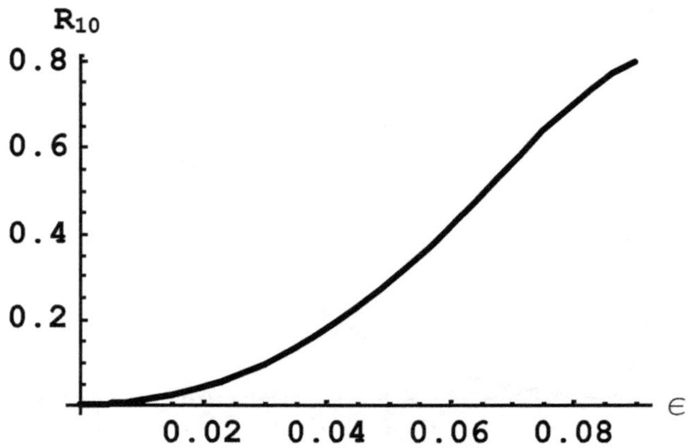

Figure 7.18. Ratio R_{10} of the intensity of the first satellite to the intensity of the main line versus the scaled dimensionless laser field ε, for the Balmer-delta line for the observation perpendicular to the laser field.

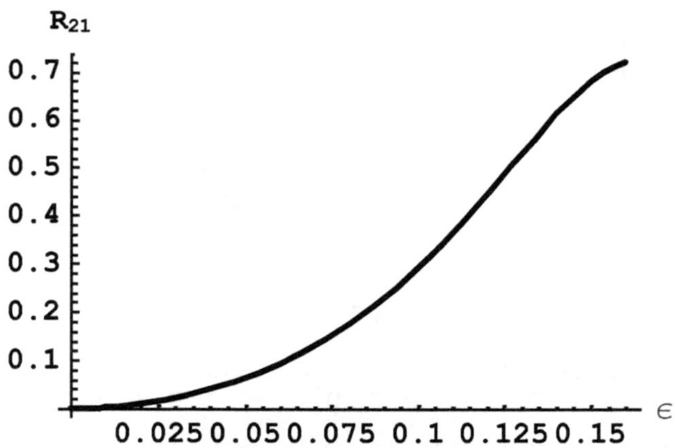

Figure 7.19. The same as in figure 7.17, but for the observation parallel to the laser field.

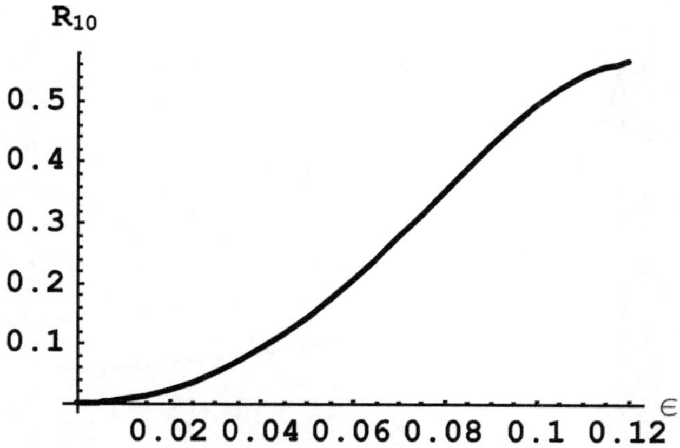

Figure 7.20. The same as in figure 7.18, but for the observation parallel to the laser field.

mechanisms. Typically, for the experimental ratio of intensities of the satellites (to each other or to the intensity of the main line) one uses the corresponding experimental ratio of *the peak intensities.*

The profile $P(\Delta\omega)$ of a spectral line affected by the optical thickness can be represented as follows:

$$P(\tau_0, \Delta\omega) = \{1 - \exp[-\tau_0 P_0(\Delta\omega)]\}/\tau_0, \tag{7.16}$$

where $P_0(\Delta\omega)$ is the profile of the absorption coefficient normalized such that $P_0(0) = 1$. At $\tau_0 = 0$, one has $P(0, \Delta\omega) = P_0(\Delta\omega)$, so that $P(0, 0) = 1$. The peak intensity of the normalized profile is

$$P(\tau_0, 0) = [1 - \exp(-\tau_0)]/\tau_0. \tag{7.17}$$

Thus, the factor reducing the peak intensity compared to the case of $\tau_0 = 0$ is

$$f(\tau_0) = [1 - \exp(-\tau_0)]/\tau_0. \tag{7.18}$$

Figure 7.21 shows the dependence of the reducing factor f on the optical depth τ_0. It is seen the greater the optical depth, the more significant becomes the reduction of the peak intensity.

Since the satellites are optically thin and $P(0, 0) = 1$, the ratio of the peak intensity of the second satellite to the peak intensity of the first satellite is still given by equation (7.6). But for the ratio of the peak intensity of the first satellite to the peak intensity of the main line we obtain

$$R_{10}(\tau_0, \varepsilon) = I(1, \varepsilon)/\,[I(0, \varepsilon)\,f(\tau_0)] \tag{7.19}$$

or

$$f(\tau_0) = I(1, \varepsilon)/\,[I(0, \varepsilon)\,R_{10}(\tau_0, \varepsilon)]. \tag{7.20}$$

Thus from the experimental ratio R_{21} one can determine the scaled amplitude of the laser field ε (as well as the laser amplitude E_0 by using equation (7.13)). Then after substituting the found value of ε and the experimental ratio R_{10} in the right-hand

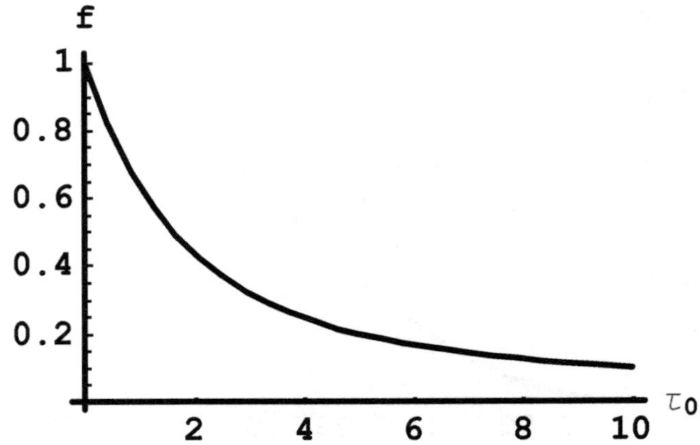

Figure 7.21. Dependence of the factor *f*, reducing the peak intensity of spectral line profiles, on the optical depth τ_0.

side of equation (7.19), one can determine the optical depth τ_0 of the main line. For the frequently used spectral lines Lyman-beta, Lyman-delta, and Balmer-beta, the ratio $I(1, \varepsilon)/I(0, \varepsilon)$ entering equations (7.19) and (7.20) can also be determined from figures 7.6, 7.8, 7.10, 7.12, 7.14, and 7.16.

In summary, we have presented a method that allows both the laser field and the opacity to be measured from the experimental spectrum of a hydrogenic line exhibiting satellites. The method is appropriate for a linearly polarized laser field at the surface of the critical density or in underdense plasma regions. We derived the necessary theoretical results analytically and showed how to use them for this purpose. This spectroscopic diagnostic should be useful for a better understanding of laser–plasma interactions, including relativistic laser–plasma interactions.

7.3 Profiles of hydrogenic spectral lines under the multimode quasimonochromatic field of the electrostatic plasma turbulence

There are a number of methods for the spectroscopic diagnostics of various types of oscillatory electric field (OEF) in plasmas—see e.g. [16]. A lot of attention has been paid to the situation where the OEF is a single-mode quasimonochromatic field: $\boldsymbol{E}(t) = \boldsymbol{E}_0\cos(\omega t)$. One practical example is a laser field at the surface of the critical density or in the underdense plasma, the spectroscopic diagnostic for which was discussed in the previous section. Another practical example is strong Langmuir waves.

However, depending on the method of the excitation of the Langmuir waves, in some situations they can be described by the model of the multimode quansimonochromatic electric field:

$$\boldsymbol{E}(t) = \sum_{j=1}^{\mathcal{N}} \boldsymbol{E}_j \cos(\omega t + \varphi_j). \tag{7.21}$$

For $\mathcal{N} \rightarrow \infty$($\mathcal{N}$ is the number of modes), the profile of a Stark component of a hydrogenic spectral line was derived in paper [17]:

$$I_{\rm L}(\Delta\omega) = \sum_{p=-\infty}^{+\infty} I_{|p|}(\tilde{\varepsilon})\exp(-\tilde{\varepsilon})\delta(\Delta\omega - p\omega), \tilde{\varepsilon} \equiv (X_{\alpha\beta}\varepsilon)^2/2. \tag{7.22}$$

In equation (7.22) $I_{|p|}(\tilde{\varepsilon})$ are the modified Bessel functions. The other notations are

$$X_{\alpha\beta} = n_\alpha q_\alpha - n_\beta q_\beta \tag{7.23}$$

and

$$\varepsilon = 3\hbar\bar{E}_0/(2m_e e\omega), \quad \bar{E}_0 = (\sum_{j=1}^{\mathcal{N}} E_j^2)^{1/2}. \tag{7.24}$$

In equation (7.23) α and β are the labels of the upper and lower Stark sublevels involved in the radiative transition, respectively, and n and q are the principal and electric quantum numbers, respectively. We recall again that the electric quantum numbers $q = n_1 - n_2$, where n_1 and n_2 are the parabolic quantum numbers.

The most interesting is the strong modulation case ($\varepsilon > 1$). In this case, the profile of a Stark component is controlled by the shape of the envelope of numerous satellites.

The hydrogenic spectral lines consist of numerous Stark components. Based on equation (7.22) for one Stark component, the corresponding Stark profiles of hydrogenic spectral lines can be calculated using formulas analogous to those from equation (7.11). The corresponding detailed tables of the calculated Stark profiles for the frequently used lines Lyman-beta, Lyman-delta, Balmer-beta, and Balmer-delta are presented in appendix C. These tables should facilitate the determination of the root-mean-square amplitude of the oscillatory field from the experimental spectral line profiles.

References

[1] Gavrilenko V P and Oks E 2006 *AIP Conf. Proc.* **874** 242

[2] Delone N B and Krainov V P 1985 *Atoms in Strong Light Fields* (Berlin: Springer)

[3] Ja B and Zel'dovich 1967 *Sov. Phys. JETP* **24** 1006

[4] Ritus V I 1967 *Sov. Phys. JETP* **24** 1041

[5] Blochinzew D I 1933 *Phys. ZZ. Sow. Union* **4** 501

[6] Gavrilenko V P and Oks E 1995 *Phys. Rev. Lett.* **74** 3796

[7] Gavrilenko V P and Oks E 1995 *J. Phys. B: At. Mol. Opt. Phys.* **28** 1433

[8] Maine P, Strickland D, Bado P, Pessot M and Mourou G 1988 *IEEE J. Quantum Electron.* **24** 398

[9] Akhiezer A I and Polovin R V 1956 *Sov. Phys. JETP* **3** 696

[10] Lünow W 1968 *Plasma Phys.* **10** 879

[11] Guerin S, Mora P, Adam J C, Heron A and Laval G 1996 *Phys. Plasmas* **3** 2693

[12] Elton R C *et al* 1997 *J. Quant. Spectrosc. Rad. Transfer* **58** 559

[13] Sauvan P, Dalimier E, Oks E, Renner O, Weber S and Riconda C 2009 *J. Phys. B: Atom. Mol. Opt. Phys.* **42** 195001
[14] Oks E 1984 *Sov. Phys. Dokl.* **29** 224
[15] Oks E 1995 *Plasma Spectroscopy: The Influence of Microwave and Laser Fields Springer Series on Atoms and Plasmas* vol 9 (Berlin: Springer)
[16] Oks E 2017 *Diagnostics of Laboratory and Astrophysical Plasmas Using Spectral Lineshapes of One-, Two-, and Three-Electron Systems* (Singapore: World Scientific)
[17] Lifshitz E V 1968 *Sov. Phys. JETP* **26** 570

Chapter 8

Enhancement of plasma-based x-ray lasers

There are several designs for lasing in the soft x-ray range. One of them employs the recombination pumping of hydrogenic ions. The ions are initially stripped of electrons by the optical field ionization [1–3]. As for any x-ray laser, the gain is controlled by the product of the oscillator strength for the lasing transition and the width of the lasing line.

When a radiating hydrogenic ion at high-frequency is subjected to a high-frequency electromagnetic radiation, the Stark width of the lasing line can be significantly diminished, as we showed in [4, 5]. In [6] we demonstrated that the gain of the x-ray laser can be substantially increased specifically by using a linearly polarized field of an optical laser. Later this idea was exploited in [7].

In this chapter, we follow [8] to briefly present the influence of the elliptically polarized field of a powerful optical laser (EPFOL) on the Stark broadening of some x-ray hydrogenic spectral lines. It is based on the results obtained in our paper [9]. The effect of the plasma microfield is taken into account via the model microfield method [10]. We demonstrate that the employment of the EPFOL makes it possible to tune the x-ray laser in a broad range of frequencies—in addition to substantially enhancing the gain.

Here, we briefly present the final results for the L_α line of Li III ($\lambda = 135$ Å) subjected to an elliptically polarized electric field $\vec{E}(t)$ of the CO_2 laser in a plasma of the electron density $N_e = 5.0 \times 10^{18}$ cm^{-3} and at the temperature $T_e = T_i = 3$ eV. These plasma parameters are similar to those from the experiment in [2].

In figure 8.1 we display the calculated Li III L_α line profile under a *linearly* polarized field $\vec{E}(t)$ for the following values of the laser field amplitude ε_0: 0 (profile 1), 3.36×10^7 V cm^{-1} (profile 2), and 4.51×10^7 V cm^{-1} (profile 4). The L_α line polarization is perpendicular to the vector $\vec{E}(t)$. As the value ε_0 increases, the profile of the absorption coefficient of the L_α line becomes narrower—in accordance with the results of our previous papers [4–6].

doi:10.1088/978-0-7503-3679-6ch8

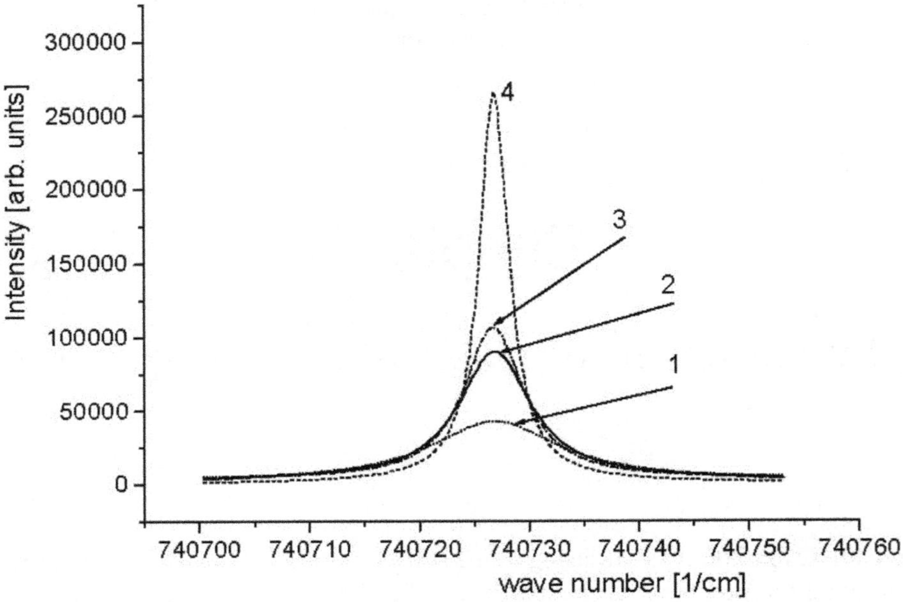

Figure 8.1. Calculated profile of the Li III L_α line (λ = 135 Å) under a *linearly* polarized electric field $\vec{E}(t)$ of the CO_2 laser in a plasma of the electron density $N_e = 5.0 \times 10^{18}$ cm^{-3} and of the temperature $T_e = T_i = 3$ eV. Perturbing ions are Li^{3+}. Profiles 1, 2, and 4 correspond to the dressing field amplitudes 0, 3.36 × 10^7 V cm^{-1}, and 4.51 × 10^7 V cm^{-1}, respectively. The x-ray radiation is polarized perpendicularly to the direction of the CO_2 laser field. Profile 3 was calculated at the dressing field amplitude 3.36 × 10^7 V cm^{-1} (like profile 2), but allowing only for the broadening by ions to illustrate the fact that, under the considered plasma parameters, the ionic contribution to the width of the profile of the absorption coefficient predominates over the electronic contribution. (Reproduced with permission from [8]. Copyright 2003 Springer.)

The ratio of halfwidths (FWHM) $\Delta\omega_{1/2}^{(k)}$ for these three profiles in figure 8.1 is

$$\Delta\omega_{1/2}^{(4)}\colon \Delta\omega_{1/2}^{(2)}\colon \Delta\omega_{1/2}^{(1)} = 0.21\colon 0.54\colon 1.00,$$

where $\Delta\omega_{1/2}^{(1)}$, $\Delta\omega_{1/2}^{(2)}$, and $\Delta\omega_{1/2}^{(4)}$ correspond to $\varepsilon_0 = 0$, $\varepsilon_0 = 3.36 \times 10^7$ V cm^{-1}, and $\varepsilon_0 = 4.51 \times 10^7$ V cm^{-1}, respectively.

Under the increase of ε_0 the profile of the absorption coefficient of the L_α line narrows and the intensity of the profile at the maximum increases (since all the profiles have the same normalization). This entails the growth of the gain of the x-ray laser under the dressing by a *linearly* polarized radiation of an optical laser, as studied earlier in [6].

Figure 8.1 also displays profile 3, calculated at $\varepsilon_0 = 3.36 \times 10^7$ V cm^{-1} (like profile 2), but allowing only for the broadening by ions. The comparison of profiles 2 and 3 illuminates the fact that, under the considered plasma parameters, the Stark broadening by ions dominates over the Stark broadening by electrons.

Figure 8.2 presents the calculated profiles of the Li III L_α in x-, y-, and z-polarizations under the CO_2 laser of $\varepsilon_0 = 3.36 \times 10^7$V cm^{-1} as the *ellipticity* degree increases from $\xi = 0$ to $\xi = 0.2$. The most instructive is the x-polarization profile.

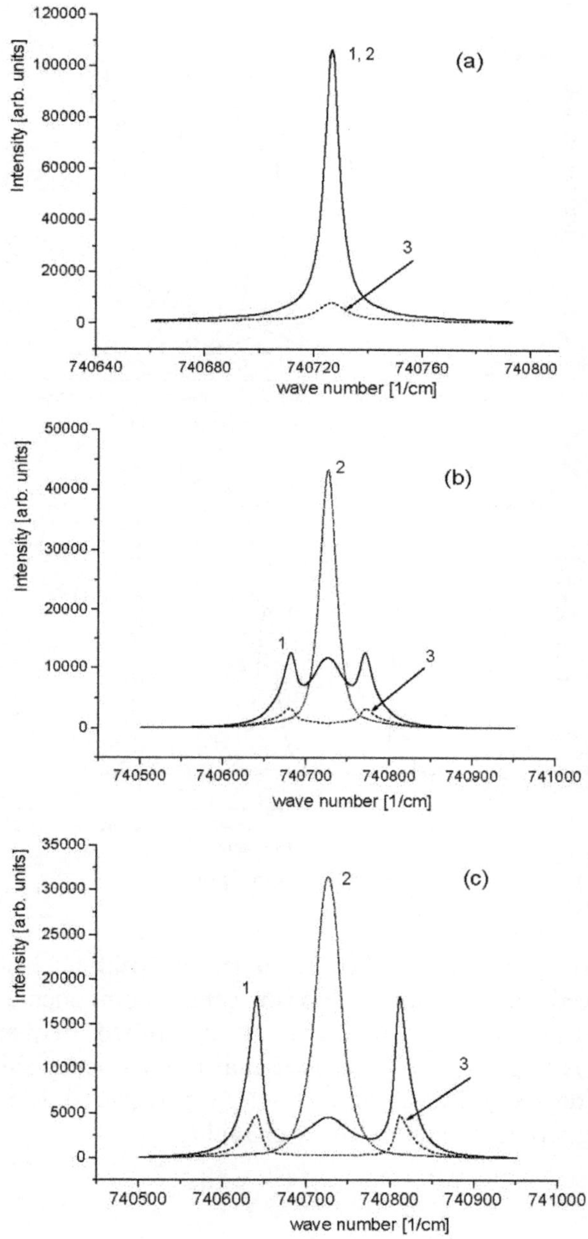

Figure 8.2. Calculated profiles of the Li III L_α line ($\lambda = 135$ Å) under an *elliptically* polarized electric field $\vec{E}(t)$ of the CO_2 laser in a plasma of the electron density $N_e = 5.0 \times 10^{18}$ cm^{-3} and of the temperature $T_e = T_i = 3$ eV. The dressing field amplitude is 3.36×10^7 V cm^{-1}. Perturbing ions are Li^{3+}. (a), (b), (c), (d), and (e) correspond to the ellipticity degrees of 0, 0.05, 0.10, 0.15, and 0.20, respectively. In each of the sub-figures, profiles 1, 2, and 3 correspond to the x-ray radiation polarized in the x-, y-, and z-directions, respectively, where z is the direction of the major axis of the ellipse and x is the direction of the minor axis of the ellipse. (Reproduced with permission from [8]. Copyright 2003 Springer.)

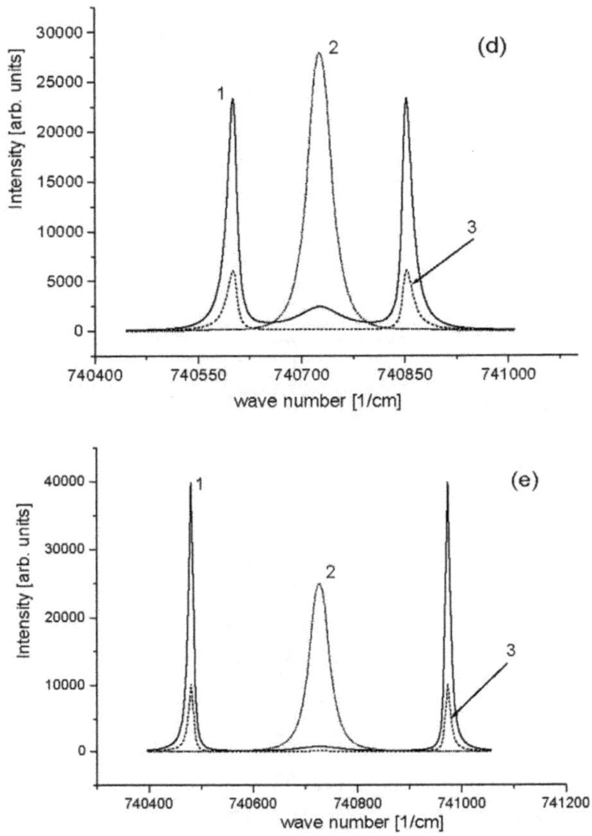

Figure 8.2. (Continued.)

It shows a splitting into three. For $\xi_{\leftrightarrow}$ 0.05, the peak intensity of the lateral components exceeds the peak intensity of the central component. Thus it should be feasible to generate the x-ray radiation at the shifted frequencies (for the x-polarization). It is important to emphasize that the frequency shift is proportional to the ellipticity degree ξ. Therefore, it should be possible to design a *tunable* x-ray laser—the tuning would be facilitated by changing ξ.

References

[1] Nagata Y, Midorikawa K, Kubodera S, Obara M, Tashiro H and Toyoda K 1993 *Phys. Rev. Lett.* **71** 3774

[2] Donnelly T D, Da Silva L, Lee R W, Mrowka S, Hofer M and Falcone R W 1996 *J. Opt. Soc. Am.* B **13** 185

[3] Korobkin D V, Nam C H, Suckewer S and Goltsov A 1996 *Phys. Rev. Lett.* **77** 5206

[4] Gavrilenko V P and Oks E 1985 *Proc. 17th Int. Conf. on Phenom. in Ionized Gases (Budapest, Hungary)* 1081

[5] Gavrilenko V P and Oks E 1989 *Proc. 19th Int. Conf. on Phenom. in Ionized Gases (Belgrade, Yugoslavia)* 354
[6] Oks E 2000 *J. Phys. B: At. Mol. Opt. Phys.* **33** L801
[7] Alexiou S 2001 *J. Quant. Spectrosc. Radiat. Transfer* **71** 139
[8] Gavrilenko V and Oks E 2004 *Eur. Phys. J.* D **28** 253
[9] Gavrilenko V and Oks E 1983 *Opt. Commun.* **46** 205
[10] Brissaud A and Frisch U 1971 *J. Quant. Spectrosc. Radiat. Transfer* **11** 1767

Chapter 9

Enhancement of generators of coherent microwave radiation (masers)

9.1 Introduction

A widely used method for the amplification and generation of tunable infra-red (IR) radiation is based on the stimulated Raman scattering (SRS) in metal vapors and in molecular gases. Compared to other nonlinear optical methods for generating IR radiation (e.g. those based on mixing electromagnetic waves), the SRS method is technically simple to realize and does not need a phase matching condition for interacting waves. Using the SRS method IR radiation has been produced in various ranges including the far IR range (40–250 μm). A further advancement into a longer wavelength range by the traditional SRS method encounters problems, one of them being caused by the typical choice of atoms (and their energy levels) employed in the standard SRS schemes. For brevity, here and below by 'atom' we mean either an atom or a molecule, unless specified to the contrary.

In this chapter we present studies of a *dipole gas* interacting with two strong electromagnetic fields: with a (laser) field of a frequency ω_L and with a second (IR or microwave) field of a much lower frequency $\omega \ll \omega_L$, under the resonance condition $\omega_L \approx \omega_{21} + q\omega$ ($q = 1, 2, 3, \ldots$), where ω_{21} is the frequency of the atomic transition between some lower state 1 and some upper state 2. By the term 'dipole gas' we mean a gaseous medium of atoms possessing permanent dipole moments d_{11} in a lower state 1 and/or d_{22} in an upper state 2. We show that while propagating in the medium the wave of the frequency ω is amplified at the expense of the pumping wave energy.

9.2 Density matrix and polarization of the medium

In this section we follow [1]. We consider some lower (1) and upper (2) states of a dipole atom under the important assumption that $d_{22} - d_{11} \neq 0$. For simplicity, we

doi:10.1088/978-0-7503-3679-6ch9

regard levels 1 and 2 to be non-degenerate and the dipole matrix element d_{12} to be real. The atom interacts with two collinear electric fields:

$$E_1(t) = a_1 \cos(\omega_L t + \phi_1), \quad \omega_L \approx \omega_{21}; \quad E_1(t) = a_1 \cos(\omega t + \phi_1), \quad \omega \ll \omega_L. \quad (9.1)$$

The equation for the density matrix ρ in the interaction representation is

$$i\hbar\partial\rho/\partial t = [V_1 + V_2, \rho] + i\hbar\gamma\rho, \quad V_1 = -dE_1(t), \quad V_2 = -dE_2(t). \quad (9.2)$$

Here γ is the relaxation operator. In the rotating wave approximation:

$$\begin{aligned} &i\hbar\partial\rho_{11}/\partial t = -V\rho_{21}\exp[i(t\Delta + \varphi_1)] - \text{c.c.} + i\hbar W_{21}\rho_{22}, \\ &i\hbar\partial\rho_{12}/\partial t = -V(\rho_{22} - \rho_{11})\exp[i(t\Delta + \varphi_1)] + \rho_{12}(d_{22} - d_{11})E_2(t) - i\hbar\rho_{12}/\tau_{12}, \quad (9.3) \\ &\rho_{22} = 1 - \rho_{11}, \quad \rho_{21} = \rho_{12}^*, \quad V = d_{12}a_1/2. \end{aligned}$$

Here $\Delta = \omega_L - \omega_{21}$ is the detuning ($|\Delta| \ll \omega_L$), W_{21} is the rate of the spontaneous emission 2→1, τ_{12}^{-1} is the transverse relaxation constant, and the symbol * denotes the complex conjugation.

The substitution is

$$\rho_{12} = \rho'_{12}\exp[iu\sin(\omega t + \varphi_2)], \quad u = (d_{11} - d_{22})a_2/(\hbar\omega). \quad (9.4)$$

We introduce the detuning from the multiquantum resonance involving both fields using the formula

$$\omega_L = \omega_{21} + q\omega - |\mu|, \quad (9.5)$$

where $|\mu| \ll \omega$ and q is an integer.

We assume that

$$|\varepsilon_p| \ll 1, \; p = 1, 2, 3, 4; \; \varepsilon_{1s} = V/(\hbar\omega), \; \varepsilon_2 = \mu/\omega, \; \varepsilon_3 = W_{21}/\omega, \; \varepsilon_4 = (\omega\tau_{12})^{-1}. \; (9.6)$$

Below is the solution of the system (9.3) under the conditions (9.5) obtained via applying the averaging method of Krylov–Bogoliubov–Mitropolskii [2, 3] in its second approximation. For the steady-state regime we obtain the following for quantities x_1, x_2, and x_3, whose relations to the matrix elements of ρ will be given several later:

$$\begin{aligned} &x_1 = y_1 + [2V/(\hbar\omega)] \sum_{p=-\infty,p\neq 0}^{\infty} p^{-1}\exp[ip(\omega t + \varphi_2)][y_2 B_{q,p}(u) + iy_3 A_{q,p}(u)], \\ &x_2 = y_2 - [V/(2\hbar\omega)]y_1 \sum_{p=-\infty,p\neq 0}^{\infty} p^{-1}\exp[ip(\omega t + \varphi_2)]B_{q,p}(u), \quad (9.7) \\ &x_3 = y_3 - i[V/(2\hbar\omega)]y_1 \sum_{p=-\infty,p\neq 0}^{\infty} p^{-1}\exp[ip(\omega t + \varphi_2)]A_{q,p}(u), \end{aligned}$$

where

$$\xi_1 = -(1 + \mu_u^2\tau_{12}^2)/f_q, \qquad \xi_2 = VJ_q(u)\mu_u\tau_{12}/(\hbar f_q),$$
$$\xi_3 = -VJ_q(u)\tau_{12}/(\hbar f_q),$$
$$Aq, p(u) = Jq + p(u) + Jq - p(u),$$
$$Bq, p = Jq + p(u) - Jq - p(u). \tag{9.8}$$

Here $J_p(x)$ are the Bessel functions. The denominators f_q in equation (9.8) have the form

$$1 + \mu_u^2\tau_{12}^2 + 4V^2J_q^2(u)\tau_{12}/(\hbar^2 W_{21}). \tag{9.10}$$

The third term in equation (9.10) represents an unusual, complicated type of saturation where the parameters of both fields are entangled. The second term in equation (9.10) can be treated as a correction (caused by finite values of both fields) to the approximate resonance condition $q\omega = \Delta$ (the latter would be valid for vanishingly small fields). The quantity μ_u in equation (9.10) is defined as follows:

$$\mu_u = \mu - 4V^2S_q(u)/(\hbar^2\omega),\ S_q(u) = \sum_{r=1}^{\infty}\left[J_{q-r}{}^2(u) - J_{q+r}{}^2(u)\right]/(2r). \tag{9.11}$$

Equation (9.7) provides the solution for the system (9.3). The diagonal matrix elements of the operator ρ are

$$\rho_{11} = (1 - x_1)/2, \qquad \rho_{22} = (1 + x_1)/2. \tag{9.12}$$

The nondiagonal element ρ_{12} is

$$\rho_{11} = (x_2 + \mathrm{i}x_3)\exp\{\mathrm{i}[u\sin(\omega t + \varphi_2) - \mu t - q\varphi_2 - \varphi_1]\}. \tag{9.13}$$

The above analytical solution is valid for a coarse-grained time scale $t \gg \max(W_{21}^{-1}, \tau_{12})$. This solution allows us to calculate the polarization of the medium

$$P = N\,\mathrm{Tr}(\rho D). \tag{9.14}$$

Here D is the operator of the atomic dipole moment and N is the atomic density. The polarization P exhibits two types of oscillating terms P_{I} and P_{II}:

$$P_{\mathrm{I}} = \exp[\mathrm{i}(\omega_{\mathrm{L}}t + \varphi_1)]\sum_{r=-\infty}^{\infty} P(\omega_{\mathrm{L}} + r\omega)\exp[\mathrm{i}r(\omega t + \varphi_2)] + \mathrm{c.c.},$$
$$P_{\mathrm{II}} = \sum_{k=-\infty, k\neq 0}^{\infty} P(k\omega)\exp[\mathrm{i}k(\omega t + \varphi_2)]. \tag{9.15}$$

In particular, imaginary P'' parts of the coefficients $P(\omega_{\mathrm{L}} + r\omega)$ and $P(k\omega)$ are found to be as follows:

$$P''(\omega_L + r\omega) = -d_{12}^2 a_1 N\tau_{12} J_q(u) J_{q+r}(u)/(2\hbar f_q),$$
$$P''(k\omega) = -d_{12}^2(d_{22} - d_{11}) a_1^2 N\tau_{12} J_q(u)\big[J_{q+k}(u) - J_{q+k}(u)\big]/(4\hbar^2 k\omega f_q). \tag{9.16}$$

9.3 Amplification of a microwave or infra-red radiation by the stimulated scattering in a dipole gas

In this section we follow [4]. Both waves of frequencies $\omega_L = \omega_1$ and $\omega = \omega_2$ propagate along the axis z in a dipole gas in the halfspace of $z \geqslant 0$ under a multiquantum resonance:

$$\omega_1 = \omega_{21} q\omega_2 - \mu, \tag{9.17}$$

where μ is the detuning. The idea of the amplification of the lower frequency wave in a dipole gas was put forward in [5, 6]. Here we show that it is the resonance condition (9.17) (which is the same as (9.5)) under which the lower frequency wave is amplified taking the energy from the pumping optical laser wave.

The Maxwell equations are

$$\mathbf{E}(\mathbf{r}, t) = (4\pi/c^2)\mathbf{P}_{tt}(\mathbf{r}, t). \tag{9.18}$$

The total electric field is

$$E(r, t) = \sum_{s=1}^{2}[\varepsilon_s(k_s, \omega_s) + \text{c.c.}], \tag{9.19}$$

where

$$\varepsilon_s(k_s, \omega_s) = 2^{-1}\mathbf{E}_{0s}(z)\ \exp[\mathrm{i}(\omega_s t - k_s z + \phi_{0s})]. \tag{9.20}$$

Here $E_{01}(z)$ and $E_{02}(z)$ are amplitudes of the waves and k_1 and k_2 are the corresponding wave vectors.

In analogy to equation (9.18) the polarization of the gas is

$$P(r, t) = \sum_{s=1}^{2}[\eta_s \varepsilon_s(k_s, \omega_s) + \text{c.c.}]. \tag{9.21}$$

For the rest of this chapter the imaginary parts η_s'' of the coefficients η_s ($s = 1, 2$) are important. The expressions for them can be derived from equation (9.16) by the substitution $a_1 = |E_{01}|$, $a_2 = |E_{02}|$, $r = 0$, $k = 1$:

$$\eta_1'' = -d_{12}^2 a_1 N\tau_{12}\big[J_q(u)\big]^2/(2\hbar f_q), \quad \eta_2'' = -q\eta_1''\,|E_{01}/E_{02}|^2,$$
$$u \equiv (d_{11} - d_{22})|E_{02}|/(\hbar\omega_2). \tag{9.22}$$

Here

$$f_q = 1 + \mu_u{}^2\tau_{12}{}^2 + 4V^2 J_q{}^2(u)\tau_{12}/(\hbar^2 W_{21}),$$
$$V \equiv d_{21}|E_{01}|/2. \qquad (9.23)$$

Substituting equations (9.20)–(9.23) in equation (9.22) we obtain

$$(\mathrm{d}^2E_{0s}/\mathrm{d}z^2) - 2ik_s(\mathrm{d}E_{0s}/\mathrm{d}z) + 4\pi\eta_s k_s{}^2 E_{0s} = 0, \quad s = 1, 2. \qquad (9.24)$$

We utilize a slowly varying amplitude approximation (see e.g. [7, 8]). It is valid under the requirement

$$|k_s(\mathrm{d}E_{0s}/\mathrm{d}z)| \gg |\mathrm{d}^2E_{0s}/\mathrm{d}z^2|, \quad s = 1, 2. \qquad (9.25)$$

The system (9.24) simplifies to

$$-\mathrm{i}(\mathrm{d}E_{0s}/\mathrm{d}z) + 2\pi k_s\eta_s E_{0s} = 1, \quad s = 1, 2. \qquad (9.26)$$

A conservation law can be derived from system (9.26):

$$|E_{01}(z)|^2/\omega_1 + |E_{02}(z)|^2/(q\omega_2) = K,$$
$$K \equiv |E_{01}(0)|^2/\omega_1 + |E_{02}(0)|^2/(q\omega_2). \qquad (9.27)$$

The physical meaning is that the absorption of one quantum of frequency ω_1 is always accompanied by the emission of q quanta of the frequency ω_2.

Using equations (9.26) and (9.27) we obtain the differential equation determining the absolute value of the amplitude $|E_{02}|$ versus z:

$$(\mathrm{d}|E_{02}|/\mathrm{d}z) - 2\pi k_2\eta_2''|E_{02}| = 0. \qquad (9.28)$$

The solution of equation (9.28) has the form of a one-fold integral

$$z = (2\pi k_2)^{-1}\int(\eta_2''|E_{02}|)^{-1}d|E_{02}|, \qquad (9.29)$$

and it is possible to eliminate the dependence of η_2'' on $|E_{01}|$ by employing the conservation law (9.27).

For $q = 1$, under the assumption

$$|u| \ll 1, \qquad (9.30)$$

the quantity η_2'' practically does not depend on E_{02}. Thus

$$|E_{02}(z)| = |E_{02}(0)| \exp(2\pi k_2\eta_2'' z). \qquad (9.31)$$

Equation (9.31) shows the exponential growth of $|E_{02}(z)|$ in this case.

However, as $|E_{02}(z)|$ grows (and consequently with the decrease of $|E_{01}(z)|$ according to equation (9.27)), the quantity η_2'' starts decreasing. This entails the retardation of the growth of I$|E_{02}(z)|$.

In the saturation, equation (9.22) for η_2'' simplifies to

$$\eta_2'' \approx q\hbar N W_{21}/(2\,|E_{02}|^2). \qquad (9.32)$$

Substituting equation (9.32) in equation (9.29) we obtain

$$|E_{02}(z)| \approx (2\pi q\hbar N\omega_2 W_{21}z/c)^{1/2}. \tag{9.33}$$

Below are the results of numerical calculations of the amplification for cesium vapor under an electric field $F = 25$ kV/cm. The states 1 and 2 are 6^2s and 12^2P, respectively. The wavelength of the transition 6^2s–12^2P is $\lambda = 335$ nm (see e.g. tables in [9]). The field F intermixes states of close levels 12^2P and 11^2D separated by approximately 40 cm^{-1} (see e.g. tables in [9]), yielding the permanent dipole moment of the 12^2P state.

Figure 9.1 depicts the dependence $|E_{02}(z)|$ for frequencies $\omega_2 = 8$ cm^{-1} (the lower, solid line) and $\omega_2 = 16$ cm^{-1} (the upper, dotted line) under the resonance (9.17) at $\mu = 0$, $q = 1$, $W_{21} = 3 \times 10^{-6}$ cm^{-1}, $\tau_{12}^{-1} = 6 \times 10^{-2}$ cm^{-1}. The input lower frequency field is $|E_{02}(0)| = 1$ V cm^{-1}. The input pumping radiation has the intensity $P_I(0) = 7.7 \times 10^8$ W cm^{-2}.

For input field strengths corresponding to $|u| > 1$, the most effective amplification could occur at $q > 1$. Figure 9.2 shows the character of growth of $|E_{02}|$ in the same dipole gas as in figure 9.1, but for the resonance involving two quanta of the lower frequency field ($q = 2$). The curves in figure 9.2 correspond to a higher input strength compared to figure 9.1, $|E_{02}(0)| = 300$ V cm^{-1}.

9.4 Extension of lasing media and lasing schemes

In this section, instead of a system of two levels with permanent dipole moments dressed by a resonant laser field, we consider a system of three levels without

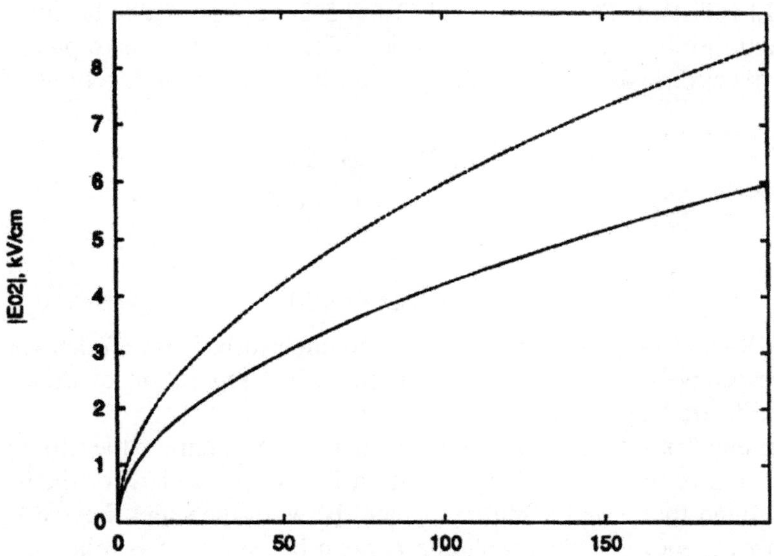

Figure 9.1. Calculated dependence $|E_{02}(z)|$ for two different frequencies $\omega_2 = 8$ cm^{-1} (the lower, solid line) and $\omega_2 = 16$ cm^{-1} (the upper, dotted line) under the resonance (9.17) at $\mu = 0$, $q = 1$. Reproduced with permission from [4]. Both curves correspond to a very small input lower frequency field: $|E_{02}(0)| = 1$ V cm^{-1}. The dipole gas consists of Cs vapor of density 5×10^{18} cm^{-3} placed in a static electric field $f = 25$ kV cm^{-1} (see the text).

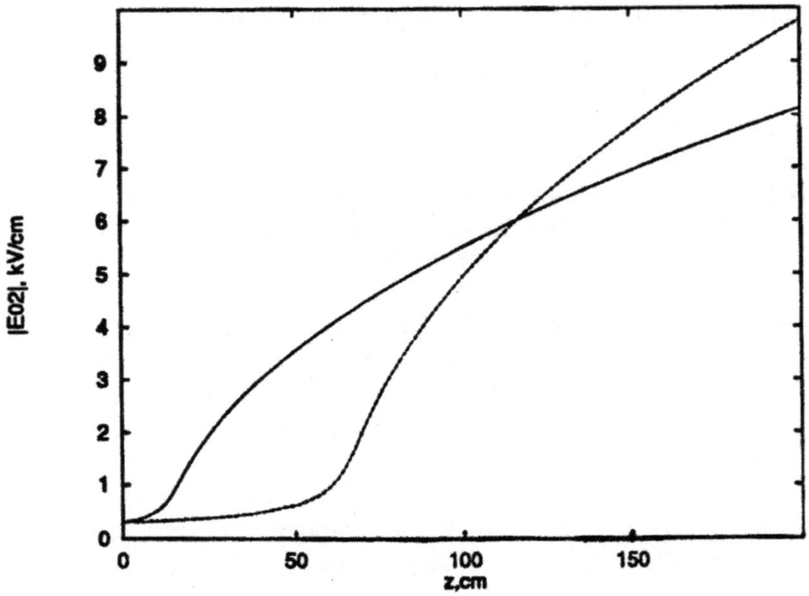

Figure 9.2. The same as figure 9.1, but for the resonance involving two quanta of a lower frequency field ($q = 2$). Reproduced with permission from [4]. Curves are calculated for a higher input strength compared to figure 9.1, $|E_{02}(0)| = 300$ V cm^{-1}. At relatively small z, the higher output is for the frequency $\omega_2 = 8$ cm^{-1} (compared to the output for the frequency $\omega_2 = 16$ cm^{-1}), but at relatively large z the situation reverses.

permanent dipole moments: the two levels 2 and 3 are relatively close to each other and level 1 is distant. The transitions 1–2 and 2–3 are supposed to be dipole-allowed, while the transition 1–3 is dipole-forbidden. The central point is to place the system into a static electric field F that will significantly mix up wave functions of the states 2 and 3:

$$\begin{aligned} \psi_2 &= \varphi_2 \cos(B/2) - \varphi_3 \sin(B/2), \\ \psi_3 &= \varphi_3 \cos(B/2) - \varphi_2 \sin(B/2), \end{aligned} \tag{9.34}$$

where

$$\tan B = 2z_{32}F/\omega_{32}(0). \tag{9.35}$$

Here, Ψ_2, Ψ_3 and ψ_2, ψ_3 are the perturbed and unperturbed wave functions of states 2 and 3, respectively, and $\omega_{32}(0)$ is the unperturbed separation of these levels; the atomic units are employed.

A dressing laser field in this scheme may be resonant either to the allowed transition 1–2 or to the forbidden transition 1–3. In particular, for the former case, we have found that a dipole matrix element between the states dressed by the laser field of a frequency and the amplitude E_{01} can be expressed as follows:

$$\begin{aligned} &\langle \chi_1 | z | \chi_2 \rangle\ \Omega^{-1} |V| z_{32} \sin B, \\ &\Omega = (\Delta^2 + 4V^2)^{1/2}, \quad \Delta = \omega_{\rm L} - \omega_{21}(F), \\ &V = \langle \varphi_1 | z | \psi_2 \rangle\ E_{0L}/2 = (z_{12}E_{0L}/2)\cos(B/2). \end{aligned} \tag{9.36}$$

Thus at $F > 0$ the dressed states are coupled by a non-zero matrix element. Therefore, a microwave amplification/generation is possible at frequencies $\Omega(F)/k$, where $k = 1, 2, 3, \ldots$.

One of the feasible practical schemes for the realization of this principle of lasing seems to be the following. We consider a beam of lithium atoms crossing a static electric field *F*. A laser tuned to the transition 2S–2P of Li ($\lambda_2 = 460.3$ nm) populates the level 2P—see figure 9.3. A second laser tuned to the transition 2P–4D of Li forms dressed atomic states at the latter transition ($\lambda_2 = 460.3$ nm). It is important to emphasize that both lasers are in the visible range, for which plenty of commercial types of lasers can be used.

We found out that for positive detunings from the resonance, the lithium atoms in a static electric field will amplify the microwave radiation at frequencies of the order of the Rabi frequency—more specifically, at frequencies $\Omega(F)/k$, where $k = 1, 2, 3, \ldots$.

The efficiency of the amplification of the microwave radiation can be enhanced by increasing absolute populations of dressed states of a lithium atom. The latter can be achieved in a regime where one laser beam of the frequency ω_{L1} saturates the transition 2S–2P and another laser beam of the frequency ω_{L2} saturates the transition 2P–4D. In other words, in this case both laser fields should be intensive —see figure 9.4.

We emphasize the following important feature of dressed atomic levels in the case under consideration. There are three levels (2S, 2P, and 4D) of a lithium atom interacting with two strong laser fields.

Therefore the energy levels of a dressed atom represent triplets (see figure 9.5) rather than doublets, as was the case for a resonant interaction of a strong laser field with a two-level system.

For the experimental realization of the amplification of microwave radiation driven by a powerful laser field, one of the best candidates is a gas consisting of dipole molecule CS. This could be similar, though not identical, to the experiments

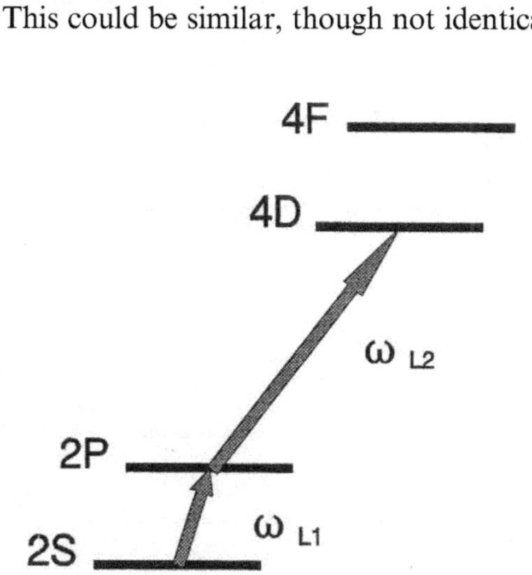

Figure 9.3. Scheme of pumping of the level 4D of lithium with two lasers of the visible range.

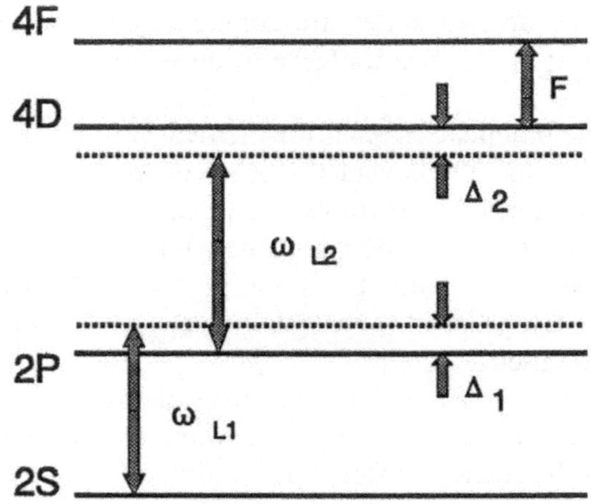

Figure 9.4. Scheme of a resonant interaction of a lithium atom with two strong laser fields ω_{L1} and ω_{L2}, as well as with a static electric field F that mixes up states of the levels 4D and 4F. Here Δ_1 and Δ_2 are the detunings of the resonances with the laser fields ω_{L1} and ω_{L2} at the transitions 2S–2P and 2P–4D, respectively.

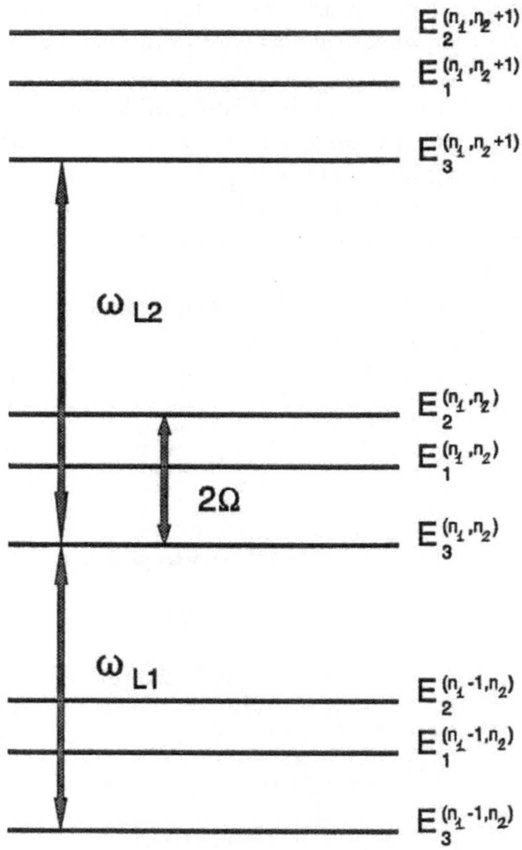

Figure 9.5. Energy levels of a lithium atom dressed by two laser fields ω_{L1} and ω_{L2}. The induced transitions between close levels $E_p^{(n1,\,n2)}$–$E_{p'}^{(n1,\,n2)}$ can cause the amplification of the microwave radiation.

reported in [10], where laser radiation was tuned to the frequency ν_{21} of the transitions between some rotational sublevels of the ground ($X^1\Sigma^+$) and excited ($A^1\Pi$) electronic states of the CS molecules—see figure 9.6.

Our main idea is—in the experimental set-up from [10]—to 'substitute' the static electric field by a microwave field. For the amplification of the microwave radiation of a frequency ν_M, the laser radiation should be tuned to a frequency

$$\nu_L = \nu_{21} + \nu_M. \tag{9.37}$$

A collateral benefit of using a dipole gas medium is the opportunity for an original spectroscopic *in situ* diagnostics of the effect of amplification of a microwave field by the laser-induced fluorescence of the molecules. We consider two possible diagnostic schemes.

The gist of the first scheme may be explained by the example of CS molecules as follows. Each rotational level of the excited Π-state is split into two close sublevels of opposite parities (Λ-doubling sublevels). It is important that these sublevels are coupled by a non-zero dipole matrix element. We tune the laser to a frequency of R-branch. If there is no other electric field, the laser radiation would populate only one sublevel of the doublet. Therefore, in the laser-induced fluorescence spectrum, in addition to the R-branch component, the P-branch component would show up, but

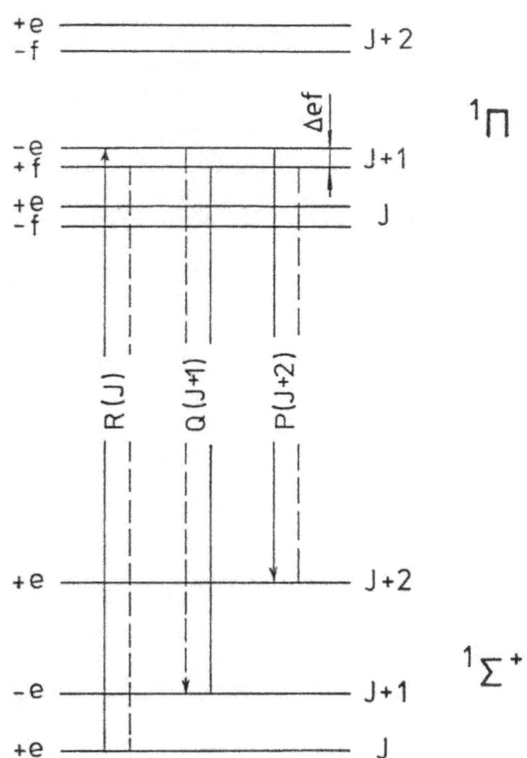

Figure 9.6. Transitions between rotational sublevels of the ground and excited states of the CS molecules (see the text).

the Q-branch (forbidden) component will remain of zero intensity. However, under an additional electric field (static or microwave) states of the doublet will be intermixed. Consequently, in the LIF spectrum, all three components will show up including the Q-branch component. The intensity ratio of components of Q- and P-branches can be used for local diagnostics of the static or microwave fields [10–13].

However, our analysis shows that this scheme will not be sensitive enough for conditions of the proposed microwave amplification experiment in a CS gas. Indeed, for a sufficient sensitivity the following condition should be met:

$$2dE_0/(\hbar\nu_M) \equiv a \geqslant 1, \tag{9.38}$$

where d is the dipole matrix element between the states of the doublet and E_0 is the microwave frequency. For microwave frequencies $\nu_M = 30$–100 GHz and reasonable amplitudes $E_0 < 1$ kV/cm, calculations result in $a \ll 1$.

The second scheme is based on the following idea. Under a superposition of laser and microwave fields satisfying condition (9.37), there occur two effects in a CS gas: (i) the polarization of the medium and (ii) the pumping up of the upper 'working' level of CS molecules (level 2 in figure 9.7). The polarization of the medium for the frequency ν_M is responsible for the primary goal—for the microwave amplification. The second effect—an increase of the upper level population—can be used for diagnostics of the primary effect.

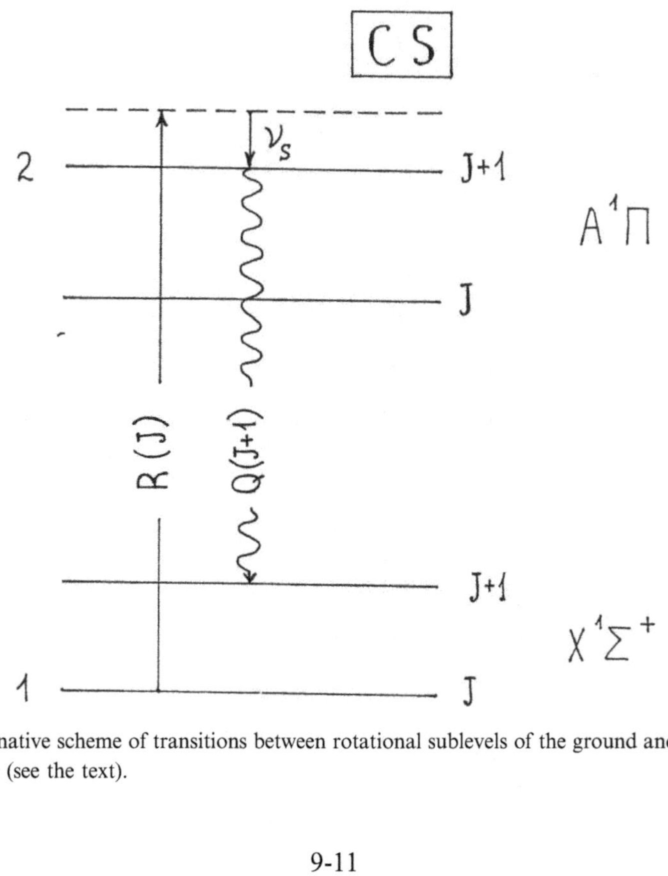

Figure 9.7. Alternative scheme of transitions between rotational sublevels of the ground and excited states of the CS molecules (see the text).

The population increase is controlled by a local strength of the microwave field. Therefore, this local microwave field can be measured by the intensity of the laser-induced fluorescence signal corresponding to a spontaneous transition from level 2 onto rotational levels of the ground electronic state $X^1\Sigma^+$ of the CS molecules.

As an example, in figure 9.7 the working levels 1 and 2 are chosen in such a way that the transition between them corresponds to the R-branch. In this case, the *in situ* diagnostics of the microwave amplification can be performed by measuring the laser-induced fluorescence signal of the Q-branch transition.

The second scheme seems to be more promising than the first one.

9.5 Consecutive cells set-up

We showed that the complex amplitude of the microwave field

$$E = v \exp(\mathrm{i}\varphi) \tag{9.39}$$

propagating in the z-direction can be found by solving the system of differential equations

$$\begin{aligned} \varphi' + 2\pi k a(v) &= 0, \\ v' - 2\pi k b(v) &= 0. \end{aligned} \tag{9.40}$$

The real (a) and imaginary (b) parts of the nonlinear susceptibility depend on v through the Bessel function $J_q(u)$. Namely, a is proportional to $J_q(u)$ and b is proportional to $J_q^2(u)$, where

$$u \equiv (ea_0/\hbar)(d_{22} - d_{11})v/\omega. \tag{9.41}$$

Here q is the number of quanta of the microwave field entering together with a laser quantum into the resonance with an atomic transition ω_{21}:

$$\omega_{\mathrm{L}} \approx \omega_{21} + q\omega. \tag{9.42}$$

The solution of the system (9.40) shows an exponential growth of the microwave field amplitude

$$v(z) = v(0) \exp[2\pi k b(0) z] \tag{9.43}$$

at the initial stage of the amplification. However, in the course of propagation, with the increase of $v(z)$ the quantity $b(v)$ starts decreasing, thus retarding the growth of $v(z)$.

It is advisable to start from the resonance (9.42) involving one quantum of this field. Then, with the microwave field growth, one should switch to the multiquantum resonance in (9.42), i.e. to $q > 1$. This can be understood by taking into account that in the system (9.40) the quantity b is proportional to $J_q^2(u)$, while $J_q^2(u)$ is proportional to u^{2q} for $|u| \ll 1$.

In addition, the system (7.2) possesses a conservation law (as we presented above)

$$|E_{\mathrm{L}}(z)|^2/\omega_{\mathrm{L}} + |E(z)|^2/(q\omega) \quad |E_{\mathrm{L}}(0)|^2/\omega_{\mathrm{L}} + |E(0)|^2/(q\omega), \tag{9.44}$$

where

$$E_L = A \exp(i\psi), \tag{9.45}$$

is a complex amplitude of the laser field. Thus it is seen that the maximum possible value of the amplified microwave field $v(z)$ increases as q grows.

Thus the best design should involve several consecutive cells containing the same dipole gas. In cell number 1, the one-quantum resonance should be utilized. Then the output from cell number 1 should be used as the input for cell number 2, where the two-quantum resonance should be used, and so on. This design would ensure the most effective amplification of the microwave radiation.

9.6 Summary and list of prospective lasing media

The above results constitute novel principles for tunable generators or amplifiers of infra-red or microwave radiation for longer wavelengths than the traditional methods can reach. No phase matching required.

The most effective amplification can be achieved by using the consecutive cells set-up. In this set-up, as the microwave signal amplifies in the cell using the one-quantum resonance with the microwave field, the switch would occur for the second cell, where the two-quantum resonance would be employed, and so on.

Here is the list of the prospective lasing media:

- Cesium vapor.
- Lithium vapor.
- CS molecules.

References

[1] Gavrilenko V P and Oks E 1995 *J. Phys. B: At. Mol. Opt. Phys.* **28** 1433

[2] Krylov N N and Bogoliubov N N 1947 *Introduction to Nonlinear Mechanics* (Princeton, NJ: Princeton University Press)

[3] Bogoliubov N N and Mitropolskii Yu M 1961 *Asymptotic Methods in the Theory of Nonlinear Oscillations* (New York: Gordon and Breach)

[4] Gavrilenko V P and Oks E 1995 *Phys. Rev. Lett.* **74** 3796

[5] Freedhoff H S 1978 *J. Phys.* B **11** 811

[6] Gavrilenko V P and Oks E A 1987 *Sov. Phys. Tech. Phys.* **32** 11

[7] Armstrong J A, Bloembergen N, Ducuing J and Pershan P S 1962 *Phys. Rev.* **127** 1918

[8] Bloembergen N 1965 *Nonlinear Optics* (New York: Benjamin)

[9] Moore C E 1958 *Atomic Energy Levels* vol 3 (Washington, DC: NSRDS)

[10] Maurmann S, Kunze H-J, Gavrilenko V and Oks E 1996 *J. Phys. B: At. Mol. Opt. Phys.* **29** 25

[11] Moore C A, Davis G P and Gottscho R A 1984 *Phys. Rev. Lett.* **52** 538

[12] Gavrilenko V P and Oks E 1986 *13th Int. Symp. on Physics of Ionized Gases (Sibenik, Yugoslavia)* (Singapore: World Scientific) p 393

[13] Oks E 1995 *Plasma Spectroscopy: The Influence of Microwave and Laser Fields* (Berlin: Springer) sect. 5.5

Chapter 10

A no-dark-energy and no-modified-gravity explanation of the dynamics of the expansion of the Universe

Currently, the two greatest mysteries in physics seem to be dark matter and dark energy. Almost all models of dark matter resort to largely unspecified never-discovered subatomic particles or introduce new physical laws—see e.g. [1–8] and references therein. One exception are papers showing the possibility that dark matter (or at least a part of it) could consist of hydrogen atoms of the second flavor, whose existence is proven by the analysis of atomic experiments and could also explain the latest observations of a 21 cm radio line from the early Universe [9, 10].

There are a lot of field and particle candidates for dark energy (again in addition to suggestions to introduce new gravitational degrees of freedom, of which massive gravity is one examples), but these candidates have not been discovered so far—see e.g. [11–14] and references therein. This is even more intriguing because the commonly accepted view is that the Universe is currently in the dark energy dominance era (estimated to have started about 5 billion years ago)—the era where yet unknown dark energy dominates over gravitation and is responsible for the observed acceleration of the expansion of the Universe.

In the current chapter, we introduce a scenario providing a possible explanation of the entire history of the expansion of the Universe: both the era of the decelerating expansion and the current era of the accelerated expansion. In this model there is no need to resort to dark energy or to new gravitational degrees of freedom (mass gravity, etc).

10.1 Description of the scenario

We present this scenario as the following sequence of logical steps:

1. We consider a 'gas' of a large number of gravitating neutral non-relativistic particles. At any instant of time the gas has a subsystem of relatively

doi:10.1088/978-0-7503-3679-6ch10

isolated pairs of particles. The subsystem is open—in the course of time, some pairs could stop qualifying as subsystem members (because they can no longer be considered as relatively isolated), while some other pairs could become relatively isolated and qualify as new members of the subsystem.

2. Pairs of the positive total energy ($E > 0$) lose energy through gravitational radiation and reach the state of $E = 0$.
3. The virial theorem applies to the motion of pairs of $E = 0$—contrary to many textbooks that state that the virial theorem applies only to bound states. As shown in [15], the boundness of a state is not the necessary condition for the applicability of the virial theorem. He derived a more general condition that is both necessary and sufficient. He specifically showed that the virial theorem applies to pairs of $E = 0$ interacting through Newton's gravity.
4. Therefore the subsystem of pairs of $E = 0$ is characterized by the following values of the average potential ($\langle U \rangle$) and kinetic ($\langle K \rangle$) energies: $\langle U \rangle = \langle K \rangle = 0$.
5. Now we apply Dirac's generalized Hamiltonian dynamics (GHD) to the above subsystem of pairs. Dirac designed the GHD for applications to quantum field theory [16–18]. The central point of Dirac's GHD is the addition of a linear combination of *constraints* to the *classical* Hamilton function. In the present book, a brief description of Dirac's GHD and its further development is given in appendix D.

   In [19] the authors demonstrated how to apply Dirac's GHD to microscopic systems of particles, such as atomic and molecular systems. Specifically, as constraints they used classical *integrals of motion* (see also [20, 21]). By doing so they showed, for example, that hydrogenic atoms/ions, being described by Dirac's GHD, exhibit a sequence of classical non-radiating states, i.e. states where the system does not emit electromagnetic radiation—it is radiationally stable. The set of energies of these classical non-radiating states turned out to exactly coincide with the corresponding quantum stationary states.

   The most important feature of classical non-radiating stable states is the following. In such a state, the radius-vector of the atomic electron $\mathbf{r}(t) = \mathbf{r}_0$ and the momentum $\mathbf{p}(t) = \mathbf{p}_0$, where $\mathbf{r}_0$ and $\mathbf{p}_0$ are some vector constants. Thus the electron velocity $\mathbf{v} = 0$, but *its momentum* $\mathbf{p}_0$ *generally differs from zero.* This should not be shocking: the momentum is a more complex physical quantity than the velocity. For instance, it is well-known that for a charge e of mass m in an electromagnetic field, characterized by a vector-potential $\mathbf{A}$, it is also possible to have $\mathbf{v} = [\mathbf{p} - e\mathbf{A}/(mc)]/m = 0$ while $\mathbf{p} = e\mathbf{A}/(mc) \neq 0$—see e.g. [22][1].

[1] Later, Dirac's GHD was successfully applied to other microscopic systems and to finding classical non-radiating states in these systems. Example are pairs of particles interacting through a modified Coulomb potential [23] and a spherical harmonic oscillator [24].

6. The way the authors of [19] applied Dirac's GHD to pairs of charges interacting through the Coulomb law can be similarly applied to pairs of neutral particles interacting through Newton's gravity. The result is an infinite number of classical non-radiating states of energies

$$E_k = -B/k^2, \quad B = \text{const} > 0, k = 1, 2, 3, \ldots, \infty. \quad (10.1)$$

Here k is the harmonic number of the primary frequency of the particle moving along an elliptical trajectory. It is convenient to denote $q = 1/k$, so that

$$E_q = -Bq^2, \quad q = 0, \ldots, 1. \quad (10.2)$$

It is obvious that q takes the infinite amount of rational numbers in the segment [0, 1] and that $E_{q=0} = 0$.
7. Thus, within Dirac's GHD, the pairs of gravitating neutral particles of energies $E > 0$ lose energy through gravitational radiation, reach the state of $E = 0$, and in this state the gravitational radiation vanishes. The subsystem of pairs of $E = 0$ is radiationally stable.
8. While the subsystem of pairs of $E = 0$ has $\langle K \rangle = 0$, the rest of the gas has $\langle K \rangle_{\text{rest}} > 0$ (the subscript 'rest' stands for the 'rest of gas'). The two subsystems interact and can exchange members. Therefore, the entire system tends to thermalize (due to the exchange of members of the two subsystems), so that for the entire system we have the average kinetic energy $\langle K \rangle < \langle K \rangle_{\text{rest}}$.
9. The average kinetic energy of the rest of the gas $\langle K \rangle_{\text{rest}}$ will continue to decrease in the course of time, so that the entire gas will have the average kinetic energy $\langle K \rangle \to 0$. This means that in the course of time, practically the entire gas would become thermodynamically equivalent to a set of non-radiating pairs of $E = 0$.
10. Here we come to the following central point. The 'freezing' of the pairs into classical states of $\mathbf{r}$ = const is equivalent to a partial inhibition of the gravitation. Indeed, without it, the gravitational radiation and the loss of the total energy would continue with 100% effectiveness and the pairs would relatively rapidly collapse (being considered classically). However, the partial inhibition of the gravitation slows down this process.
11. Thus as the share of the subsystem of pairs (in the entire system) increases in the course of time the *gravitational interaction within the entire system effectively decreases* and tends to zero.
12. For this scenario, the particles of the gas should have a practically infinite lifetime and should have zero or very little interaction with the rest of the matter. Therefore, the most probable candidate for these particles is non-relativistic neutrinos. The current technology allows for detecting only relativistic neutrinos—because slow moving neutrinos have very low probabilities of interactions with the rest of the matter. However, nothing prohibits neutrinos from moving at non-relativistic velocities.

10.2 Conclusions

We considered a 'gas' of a large number of gravitating neutral non-relativistic particles. We presented a step-by-step scenario, one of the central points of which was the application of Dirac's GHD to pairs of these particles. Another central point was that the virial theorem is valid for pairs of zero total energy interacting through Newton's gravity—contrary to many textbooks stating that the virial theorem applies only to bound states. We showed that as a result, the gravitational interaction within the entire system effectively decreases.

According to a recent study [25] there is a disparity between galaxies spinning clockwise and counterclockwise, so that the Universe as the whole rotates. If we were to take this into account, then the particles of the rotating gas would *expand at an increasing speed.*

The following point should be emphasized. The rotation of the Universe alone cannot explain why in the first 9 billion years or so the expansion of the Universe was decelerating. The entire situation can be explained by a combination of the inhibition of the gravitational interaction (described above) complemented by the rotation of the Universe. In other words, it took time for the gas to reach the above stage 10 (the 'freezing' of the gravitation). During this time gravitation was still dominating the opposing centrifugal force caused by the rotation of the Universe, so that the expansion of the Universe was slowing down. Then, after the gas reached the above stage 10, the centrifugal force caused by the rotation of the Universe started dominating the gravitation, so that the expansion of the Universe began to accelerate.

Thus in this model there is no need to resort to dark energy or to new gravitational degrees of freedom (mass gravity etc) to explain the entire history of the expansion of the Universe, including the observed acceleration of the expansion in the last 5 billion years or so.

References

[1] de Martino I, Chakrabarty S S, Cesare V, Gallo A, Ostorero L and Diaferio A 2020 *Universe* **6** 107
[2] Salucci P, Turini N and di Paolo C 2020 *Universe* **6** 118
[3] Bertone G and Hooper D 2018 *Rev. Mod. Phys.* **90** 045002
[4] Bertone G and Tait T M P 2018 *Nature* **562** 51
[5] Einasto J 2013 *Dark Matter and Cosmic Web Story* (Singapore: World Scientific)
[6] Majumdar D 2014 *Dark Matter: An Introduction* (Boca Raton, FL: CRC Press)
[7] Garrett K and Düda G 2011 *Adv. Astron.* **2011** 968283
[8] Sanders R H 2010 *The Dark Matter Problem: A Historical Perspective* (Cambridge: Cambridge University Press)
[9] Oks E 2020 *Res. Astron. Astrophys.* **20** 109
[10] Oks E 2020 *Atoms* **8** 33
[11] Frusciante N and Perenon L 2020 *Phys. Rep.* **857** 1
[12] Tawfik A N and Dahab E A E 2019 *Gravitation Cosmol.* **25** 103
[13] Brax P 2017 *Rep. Prog. Phys.* **81** 016902

[14] Joyce A, Lombriser L and Schmidt F 2016 *Annu. Rev. Nucl. Part. Sci.* **66** 95
[15] Pollard H 1964 *Bull. Am. Math. Soc.* **70** 703
[16] Dirac P A M 1950 *Canad. J. Math* **2** 129
[17] Dirac P A M 1958 *Proc. R. Soc.* A **246** 326
[18] Dirac P A M 1964 *Lectures on Quantum Mechanics* (New York: Academic) reprinted by Dover 2001
[19] Oks E and Uzer T 2002 *J. Phys. B: At. Mol. Opt. Phys.* **35** 165
[20] Oks E 2015 *Breaking Paradigms in Atomic and Molecular Physics* (Singapore: World Scientific) ch 4
[21] Oks E 2019 *Unexpected Similarities of the Universe with Atomic and Molecular Systems: What a Beautiful World* (Bristol: IOP Publishing) appendix A
[22] Landau L D and Lifshitz E M 1960 *Classical Theory of Fields* (Oxford: Pergamon)
[23] Camarena A and Oks E 2010 *Int. Rev. Atom. Mol. Phys.* **1** 143
[24] Oks E 2020 *Symmetry* **12** 1130
[25] Shamir L 2020 *Astrophys. Space Sci.* **365** 136

Appendix A

Correct expression for $\alpha(F)$ controlling the transmission coefficient $D(F)$ in equation (5.18) of chapter 5

The quantity $\alpha(F)$ is incorrectly given by Kulyagin and Taranukhin [1]. Indeed, in [2], in problem 4 after paragraph 50, the quantity ε (which was re-named as α by Kulyagin and Taranukhin) is given in atomic units as

$$\varepsilon = E/k^{1/2}, \tag{A.1}$$

where E is the energy counted from the top of the parabolic barrier (for energies below the top of the barrier, $\varepsilon = -|E|/k^{1/2}$). In other words, in the notation of Kulyagin and Taranukhin this should be

$$\alpha = -\left[U(E)/4 - U_{0p}\right]/k^{1/2}. \tag{A.2}$$

The quantity k, correctly calculated by Kulyagin and Taranukhin, is

$$k = (8/\Delta\eta^2)\left[U(E)/4 - U_{0p}\right], \qquad \Delta\eta = \eta_2 - \eta_1. \tag{A.3}$$

After substituting equation (A.3) in equation (A.2) we obtain

$$\begin{aligned}\alpha &= -\,\Delta\eta\left[U(E)/4 - U_{0p}\right]/\left\{8\left[U(E)/4 - U_{0p}\right]\right\}^{1/2}\\ &= -\,\Delta\eta\left\{\left[U(E)/4 - U_{0p}\right]/8\right\}^{1/2},\end{aligned} \tag{A.4}$$

which differs significantly from the incorrect Kulyagin–Taranukhin formula for α.

References

[1] Kulyagin R V and Taranukhin V D 1993 *Laser Phys.* **3** 644
[2] Landau L D and Lifshitz E M 1965 *Quantum Mechanics: Non-Relativistic Theory* (Oxford: Pergamon)

Appendix B

Weak field expansion of the integral in equation (5.15) of chapter 5

First we find the weak field limit of the roots $b(F)$, $a(F)$, and $-c(F)$ of the cubic equation

$$\eta^3 - 2U\eta^2/F + 4\beta_2\eta/F + 1/F = 0. \quad \text{(B.1)}$$

From a simple analysis it follows that in this limit $b \to \infty$, while a and c remain finite. Therefore the asymptotic value of b can be found as the largest root of the truncated equation

$$\eta^3 - 2U\eta^2/F + 4\beta_2\eta/F = 0. \quad \text{(B.2)}$$

Thus we obtain the first two terms of the asymptotic expansion of b in the form

$$b(F) = (2U/F)(1 - \beta_2 F/U^2). \quad \text{(B.3)}$$

The asymptotic values of the roots $a(F)$ and $-c(F)$ can be found from another truncated equation obtained by omitting the term η^3 in the cubic equation

$$-2U\eta^2/F + 4\beta_2\eta/F + 1/F = 0. \quad \text{(B.4)}$$

As a result we obtain

$$a(F) = \left[\left(\beta_2^2 + U/2\right)^{1/2} + \beta_2\right]\Big/U, \qquad c(F) = \left[\left(\beta_2^2 + U/2\right)^{1/2} - \beta_2\right]\Big/U. \quad \text{(B.5)}$$

The arguments of the elliptic integrals depend on the quantities $d = [(a + c)/(b + c)]^{1/2}$ and $j = [a/(b + c)]^{1/2}$. From the previous section it follows that in the weak field limit we have $d \to 0$ and $j \to 0$. Thus the task is to find the expansion of the elliptic integrals as $d \to 0$ and $j \to 0$.

The bracket containing the elliptic integrals of the first kind can be expressed as follows:

$$[\mathbf{K}(1/d^2) - \mathbf{F}(\arcsin d,\ 1/d^2)] = -id\mathbf{K}(1 - d^2). \tag{B.6}$$

Let us temporarily denote the argument of the **K**-integral in the right side as k: $k = 1 - d^2$. By introducing $k' = (1 - k^2)$ and noting that in the limit $d \to 0$ it can be approximated as $k' = 2^{1/2}d \ll 1$, we can use the asymptotic result $\mathbf{K}(k') = \ln(4/k')$ and obtain

$$[\mathbf{K}(1/d^2) - \mathbf{F}(\arcsin d,\ 1/d^2)]_{as} = -id\ln(2^{3/2}/d). \tag{B.7}$$

The bracket containing the elliptic integrals of the second kind can be represented in the form

$$[\mathbf{E}(1/d^2) - \mathbf{E}(\arcsin d,\ 1/d^2)] = (i/d)\, I_1(d),\qquad I_1(d) = \int_d^1 dt[(t^2 - d^2)/(1 - t^2)]^{1/2}. \tag{B.8}$$

By expanding $I_1(d)$ in the Taylor series we obtain

$$I_1(d) = I_1(0) + dI_1'(0) = 1 - d^2 I_2(d),\qquad I_2(d) = \int_d^1 dt[(t^2 - d^2)(1 - t^2)]^{-1/2} \tag{B.9}$$

(here $I_1'(0)$ is the derivative of $I_1(z)$). To obtain the asymptotic value of $I_2(d)$ we note that

$$[\mathbf{K}(1/d^2) - \mathbf{F}(\arcsin d,\ 1/d^2)] = -idI_2(d). \tag{B.10}$$

By comparing (B.10) and (B.6) we find

$$I_2(d) = \mathbf{K}(1 - d^2). \tag{B.11}$$

Then, by using the asymptotic value of $\mathbf{K}(1 - d^2)$ from (B.7), we obtain

$$[\mathbf{E}(1/d^2) - \mathbf{E}(\arcsin d,\ 1/d^2)]_{as} = \mathrm{i}[1/d - d\ln(2^{3/2}/d)]. \tag{B.12}$$

The bracket containing the elliptic integrals of the third kind can be represented in the form

$$[\mathbf{\Pi}(1/(d^2 - j^2),\ 1/d^2) - \mathbf{\Pi}(1/(d^2 - j^2),\ \arcsin d,\ 1/d^2)] = \mathrm{i}d(d^2 - j^2)\int_d^1 dt(t^2 - d^2 + j^2)^{-1}[(t^2 - d^2)(1 - t^2)]^{-1/2}. \tag{B.13}$$

The integrand in (B.13) has singularities at $t = d$ and at $t = 1$, where it goes to positive infinity. Inside the integration range the integrand has a minimum at $t = 3^{1/2}/2$. The location of the minimum as well as the subsequent results is obtained in the limit $d \ll 1$ (we recall that $j < d$).

We divide the integration range into two parts: from d to $d^{P(n)}$, the corresponding integral being denoted as I_{31}, and from $d^{P(n)}$ to 1, the corresponding integral being denoted as I_{32}. Here $P(n) = 1/2^n$, where $n = 1, 2, 3, \ldots$, so that $d < d^{P(n)} < 1$. The integral I_{31} can be approximated as follows:

$$I_{31} = \int_d^{P(n)} \mathrm{d}t\, j^{-2}(t^2 - d^2)^{-1/2}. \tag{B.14}$$

After integrating I_{31} from (B.14) analytically and using $n \gg 1$ we obtain

$$I_{31} = j^{-2} \ln(2/d). \tag{B.15}$$

The integral I_{32} can be approximated as follows:

$$I_{32} = \int_{P(n)}^{1} \mathrm{d}t\, t^{-3}(1 - t^2)^{-1/2}. \tag{B.16}$$

After integrating I_{32} from (B.16) analytically and using $n \gg 1$, we find that $I_{32} = \text{const} \sim 1$, so that it can be disregarded compared to $I_{31} = j^{-2} \ln(2/d)$ since $j < d \ll 1$.

Thus we obtain the asymptotic expression for the right side of (B.13) as

$$\begin{aligned}&[\boldsymbol{\Pi}(1/(d^2 - j^2),\ 1/d^2) - \boldsymbol{\Pi}(1/(d^2 - j^2),\ \arcsin d,\ 1/d^2)]_{\text{as}} \\ &= \mathrm{i}d(d^2 - j^2)j^{-2} \ln(2/d).\end{aligned} \tag{B.17}$$

By substituting (B.7), (B.12), and (B.17) in (B.4), we obtain equation (5.16) from chapter 5 for $\text{Int}_{\text{as}}(F)$.

Appendix C

Tables of Stark profiles of hydrogenic spectral lines under the multimode quasimonochromatic electrostatic turbulence in plasmas

In this appendix, referred to in section 7.3, we present detailed tables of the Stark profiles of the hydrogenic spectral lines Lyman-beta, Lyman-delta, Balmer-beta, and Balmer-delta under the multimode quasimonochromatic electrostatic turbulence in plasmas. The Stark profiles have been calculated as described in section 7.3. They are presented here in the universal form—as the function of the scaled dimensionless detuning δ from the unperturbed frequency of the spectral line:

$$\delta = 2Z_r m_e e(\Delta\omega)/(3\hbar E_{rms}). \tag{C.1}$$

In equation (C.1) Z_r is the nuclear charge of the radiating atom or ion, m_e and e are the electron mass and charge, respectively, $\Delta\omega$ is the detuning from the unperturbed frequency of the spectral line in the frequency scale, and E_{rms} is the root-mean-square amplitude of the oscillatory field.

Below, there are two tables of Stark profiles for each spectral line. One table is for the observation perpendicular to the field, and another table is for the observation parallel to the field.

Each table consists of a set of entries of the form {δ, S}, where $S(\delta)$ is the corresponding relative intensity. Each profile $S(\delta)$ is normalized to unity. In each table, the slash \ at the end of each line, except the last lines, indicates that the table continues in the next line.

Profile of the Lyman-beta line for the observation perpendicular to the field

{0, 0.141047}, {0.01, 0.141046}, {0.02, 0.141043}, {0.03, 0.141037}, {0.04, \
0.141029}, {0.05, 0.141018}, {0.06, 0.141005}, {0.07, 0.14099}, {0.08, \
0.140972}, {0.09, 0.140952}, {0.1, 0.14093}, {0.11, 0.140905}, {0.12, \
0.140878}, {0.13, 0.140849}, {0.14, 0.140817}, {0.15, 0.140783}, {0.16, \
0.140747}, {0.17, 0.140708}, {0.18, 0.140667}, {0.19, 0.140624}, {0.2, \
0.140578}, {0.21, 0.14053}, {0.22, 0.14048}, {0.23, 0.140427}, {0.24, \

0.140372}, {0.25, 0.140315}, {0.26, 0.140256}, {0.27, 0.140194}, {0.28, \
0.14013}, {0.29, 0.140063}, {0.3, 0.139994}, {0.31, 0.139923}, {0.32, \
0.13985}, {0.33, 0.139774}, {0.34, 0.139697}, {0.35, 0.139616}, {0.36, \
0.139534}, {0.37, 0.139449}, {0.38, 0.139363}, {0.39, 0.139273}, {0.4, \
0.139182}, {0.41, 0.139088}, {0.42, 0.138993}, {0.43, 0.138894}, {0.44, \
0.138794}, {0.45, 0.138692}, {0.46, 0.138587}, {0.47, 0.13848}, {0.48, \
0.138371}, {0.49, 0.138259}, {0.5, 0.138146}, {0.51, 0.13803}, {0.52, \
0.137912}, {0.53, 0.137792}, {0.54, 0.13767}, {0.55, 0.137546}, {0.56, \
0.13742}, {0.57, 0.137291}, {0.58, 0.13716}, {0.59, 0.137027}, {0.6, \
0.136893}, {0.61, 0.136756}, {0.62, 0.136616}, {0.63, 0.136475}, {0.64, \
0.136332}, {0.65, 0.136187}, {0.66, 0.136039}, {0.67, 0.13589}, {0.68, \
0.135738}, {0.69, 0.135585}, {0.7, 0.135429}, {0.71, 0.135272}, {0.72, \
0.135112}, {0.73, 0.13495}, {0.74, 0.134787}, {0.75, 0.134621}, {0.76, \
0.134454}, {0.77, 0.134284}, {0.78, 0.134113}, {0.79, 0.13394}, {0.8, \
0.133764}, {0.81, 0.133587}, {0.82, 0.133408}, {0.83, 0.133227}, {0.84, \
0.133044}, {0.85, 0.13286}, {0.86, 0.132673}, {0.87, 0.132485}, {0.88, \
0.132294}, {0.89, 0.132102}, {0.9, 0.131908}, {0.91, 0.131712}, {0.92, \
0.131515}, {0.93, 0.131315}, {0.94, 0.131114}, {0.95, 0.130911}, {0.96, \
0.130707}, {0.97, 0.1305}, {0.98, 0.130292}, {0.99, 0.130082}, {1.0, \
0.129871}, {1.01, 0.129658}, {1.02, 0.129443}, {1.03, 0.129226}, {1.04, \
0.129008}, {1.05, 0.128788}, {1.06, 0.128567}, {1.07, 0.128343}, {1.08, \
0.128119}, {1.09, 0.127892}, {1.1, 0.127664}, {1.11, 0.127435}, {1.12, \
0.127204}, {1.13, 0.126971}, {1.14, 0.126737}, {1.15, 0.126501}, {1.16, \
0.126264}, {1.17, 0.126026}, {1.18, 0.125785}, {1.19, 0.125544}, {1.2, \
0.125301}, {1.21, 0.125056}, {1.22, 0.12481}, {1.23, 0.124563}, {1.24, \
0.124314}, {1.25, 0.124064}, {1.26, 0.123813}, {1.27, 0.12356}, {1.28, \
0.123305}, {1.29, 0.12305}, {1.3, 0.122793}, {1.31, 0.122535}, {1.32, \
0.122275}, {1.33, 0.122015}, {1.34, 0.121752}, {1.35, 0.121489}, {1.36, \
0.121225}, {1.37, 0.120959}, {1.38, 0.120692}, {1.39, 0.120424}, {1.4, \
0.120154}, {1.41, 0.119884}, {1.42, 0.119612}, {1.43, 0.11934}, {1.44, \
0.119066}, {1.45, 0.118791}, {1.46, 0.118514}, {1.47, 0.118237}, {1.48, \
0.117959}, {1.49, 0.11768}, {1.5, 0.117399}, {1.51, 0.117118}, {1.52, \
0.116835}, {1.53, 0.116552}, {1.54, 0.116267}, {1.55, 0.115982}, {1.56, \
0.115696}, {1.57, 0.115408}, {1.58, 0.11512}, {1.59, 0.114831}, {1.6, \
0.114541}, {1.61, 0.11425}, {1.62, 0.113958}, {1.63, 0.113666}, {1.64, \
0.113372}, {1.65, 0.113078}, {1.66, 0.112783}, {1.67, 0.112487}, {1.68, \
0.11219}, {1.69, 0.111892}, {1.7, 0.111594}, {1.71, 0.111295}, {1.72, \
0.110995}, {1.73, 0.110695}, {1.74, 0.110394}, {1.75, 0.110092}, {1.76, \
0.109789}, {1.77, 0.109486}, {1.78, 0.109182}, {1.79, 0.108878}, {1.8, \
0.108573}, {1.81, 0.108267}, {1.82, 0.107961}, {1.83, 0.107654}, {1.84, \
0.107347}, {1.85, 0.107039}, {1.86, 0.10673}, {1.87, 0.106421}, {1.88, \
0.106112}, {1.89, 0.105802}, {1.9, 0.105491}, {1.91, 0.10518}, {1.92, \
0.104869}, {1.93, 0.104557}, {1.94, 0.104245}, {1.95, 0.103932}, {1.96, \
0.103619}, {1.97, 0.103305}, {1.98, 0.102992}, {1.99, 0.102677}, {2.0, \
0.102363}, {2.01, 0.102048}, {2.02, 0.101733}, {2.03, 0.101417}, {2.04, \

0.101101}, {2.05, 0.100785}, {2.06, 0.100469}, {2.07, 0.100153}, {2.08, \
0.0998357}, {2.09, 0.0995186}, {2.1, 0.0992014}, {2.11, 0.0988839}, {2.12, \
0.0985663}, {2.13, 0.0982485}, {2.14, 0.0979305}, {2.15, 0.0976124}, {2.16, \
0.0972941}, {2.17, 0.0969757}, {2.18, 0.0966572}, {2.19, 0.0963386}, {2.2, \
0.0960199}, {2.21, 0.0957012}, {2.22, 0.0953823}, {2.23, 0.0950635}, {2.24, \
0.0947446}, {2.25, 0.0944256}, {2.26, 0.0941067}, {2.27, 0.0937877}, {2.28, \
0.0934688}, {2.29, 0.0931498}, {2.3, 0.092831}, {2.31, 0.0925121}, {2.32, \
0.0921933}, {2.33, 0.0918746}, {2.34, 0.0915559}, {2.35, 0.0912374}, {2.36, \
0.0909189}, {2.37, 0.0906006}, {2.38, 0.0902824}, {2.39, 0.0899643}, {2.4, \
0.0896464}, {2.41, 0.0893286}, {2.42, 0.089011}, {2.43, 0.0886936}, {2.44, \
0.0883763}, {2.45, 0.0880593}, {2.46, 0.0877424}, {2.47, 0.0874258}, {2.48, \
0.0871095}, {2.49, 0.0867933}, {2.5, 0.0864774}, {2.51, 0.0861618}, {2.52, \
0.0858465}, {2.53, 0.0855314}, {2.54, 0.0852166}, {2.55, 0.0849022}, {2.56, \
0.084588}, {2.57, 0.0842742}, {2.58, 0.0839607}, {2.59, 0.0836475}, {2.6, \
0.0833347}, {2.61, 0.0830222}, {2.62, 0.0827102}, {2.63, 0.0823985}, {2.64, \
0.0820871}, {2.65, 0.0817762}, {2.66, 0.0814657}, {2.67, 0.0811556}, {2.68, \
0.0808459}, {2.69, 0.0805367}, {2.7, 0.0802279}, {2.71, 0.0799195}, {2.72, \
0.0796116}, {2.73, 0.0793042}, {2.74, 0.0789972}, {2.75, 0.0786908}, {2.76, \
0.0783848}, {2.77, 0.0780793}, {2.78, 0.0777743}, {2.79, 0.0774699}, {2.8, \
0.0771659}, {2.81, 0.0768625}, {2.82, 0.0765597}, {2.83, 0.0762573}, {2.84, \
0.0759556}, {2.85, 0.0756544}, {2.86, 0.0753537}, {2.87, 0.0750536}, {2.88, \
0.0747542}, {2.89, 0.0744553}, {2.9, 0.074157}, {2.91, 0.0738593}, {2.92, \
0.0735622}, {2.93, 0.0732657}, {2.94, 0.0729699}, {2.95, 0.0726747}, {2.96, \
0.0723801}, {2.97, 0.0720861}, {2.98, 0.0717928}, {2.99, 0.0715002}, {3.0, \
0.0712082}, {3.01, 0.0709169}, {3.02, 0.0706263}, {3.03, 0.0703363}, {3.04, \
0.0700471}, {3.05, 0.0697585}, {3.06, 0.0694706}, {3.07, 0.0691834}, {3.08, \
0.0688969}, {3.09, 0.0686112}, {3.1, 0.0683261}, {3.11, 0.0680418}, {3.12, \
0.0677582}, {3.13, 0.0674753}, {3.14, 0.0671932}, {3.15, 0.0669118}, {3.16, \
0.0666312}, {3.17, 0.0663513}, {3.18, 0.0660722}, {3.19, 0.0657938}, {3.2, \
0.0655162}, {3.21, 0.0652394}, {3.22, 0.0649633}, {3.23, 0.064688}, {3.24, \
0.0644135}, {3.25, 0.0641398}, {3.26, 0.0638669}, {3.27, 0.0635947}, {3.28, \
0.0633234}, {3.29, 0.0630529}, {3.3, 0.0627832}, {3.31, 0.0625142}, {3.32, \
0.0622461}, {3.33, 0.0619788}, {3.34, 0.0617124}, {3.35, 0.0614467}, {3.36, \
0.0611819}, {3.37, 0.0609179}, {3.38, 0.0606547}, {3.39, 0.0603924}, {3.4, \
0.0601309}, {3.41, 0.0598703}, {3.42, 0.0596105}, {3.43, 0.0593515}, {3.44, \
0.0590934}, {3.45, 0.0588361}, {3.46, 0.0585797}, {3.47, 0.0583241}, {3.48, \
0.0580694}, {3.49, 0.0578156}, {3.5, 0.0575626}, {3.51, 0.0573105}, {3.52, \
0.0570592}, {3.53, 0.0568088}, {3.54, 0.0565593}, {3.55, 0.0563106}, {3.56, \
0.0560629}, {3.57, 0.055816}, {3.58, 0.0555699}, {3.59, 0.0553248}, {3.6, \
0.0550805}, {3.61, 0.0548371}, {3.62, 0.0545946}, {3.63, 0.054353}, {3.64, \
0.0541122}, {3.65, 0.0538724}, {3.66, 0.0536334}, {3.67, 0.0533953}, {3.68, \
0.0531581}, {3.69, 0.0529218}, {3.7, 0.0526864}, {3.71, 0.0524518}, {3.72, \
0.0522182}, {3.73, 0.0519854}, {3.74, 0.0517535}, {3.75, 0.0515226}, {3.76, \
0.0512925}, {3.77, 0.0510633}, {3.78, 0.050835}, {3.79, 0.0506076}, {3.8, \
0.0503811}, {3.81, 0.0501555}, {3.82, 0.0499308}, {3.83, 0.0497069}, {3.84, \

0.049484}, {3.85, 0.0492619}, {3.86, 0.0490408}, {3.87, 0.0488205}, {3.88, \
0.0486012}, {3.89, 0.0483827}, {3.9, 0.0481651}, {3.91, 0.0479484}, {3.92, \
0.0477326}, {3.93, 0.0475177}, {3.94, 0.0473037}, {3.95, 0.0470906}, {3.96, \
0.0468783}, {3.97, 0.046667}, {3.98, 0.0464565}, {3.99, 0.0462469}, {4.0, \
0.0460382}, {4.01, 0.0458304}, {4.02, 0.0456235}, {4.03, 0.0454174}, {4.04, \
0.0452122}, {4.05, 0.0450079}, {4.06, 0.0448045}, {4.07, 0.044602}, {4.08, \
0.0444003}, {4.09, 0.0441996}, {4.1, 0.0439996}, {4.11, 0.0438006}, {4.12, \
0.0436024}, {4.13, 0.0434051}, {4.14, 0.0432087}, {4.15, 0.0430131}, {4.16, \
0.0428184}, {4.17, 0.0426246}, {4.18, 0.0424316}, {4.19, 0.0422395}, {4.2, \
0.0420483}, {4.21, 0.0418579}, {4.22, 0.0416683}, {4.23, 0.0414796}, {4.24, \
0.0412918}, {4.25, 0.0411048}, {4.26, 0.0409186}, {4.27, 0.0407333}, {4.28, \
0.0405489}, {4.29, 0.0403653}, {4.3, 0.0401825}, {4.31, 0.0400005}, {4.32, \
0.0398194}, {4.33, 0.0396392}, {4.34, 0.0394597}, {4.35, 0.0392811}, {4.36, \
0.0391033}, {4.37, 0.0389263}, {4.38, 0.0387502}, {4.39, 0.0385749}, {4.4, \
0.0384004}, {4.41, 0.0382267}, {4.42, 0.0380538}, {4.43, 0.0378817}, {4.44, \
0.0377104}, {4.45, 0.03754}, {4.46, 0.0373703}, {4.47, 0.0372015}, {4.48, \
0.0370334}, {4.49, 0.0368661}, {4.5, 0.0366997}, {4.51, 0.036534}, {4.52, \
0.0363691}, {4.53, 0.036205}, {4.54, 0.0360416}, {4.55, 0.0358791}, {4.56, \
0.0357173}, {4.57, 0.0355563}, {4.58, 0.035396}, {4.59, 0.0352366}, {4.6, \
0.0350779}, {4.61, 0.0349199}, {4.62, 0.0347628}, {4.63, 0.0346063}, {4.64, \
0.0344507}, {4.65, 0.0342958}, {4.66, 0.0341416}, {4.67, 0.0339882}, {4.68, \
0.0338355}, {4.69, 0.0336835}, {4.7, 0.0335323}, {4.71, 0.0333819}, {4.72, \
0.0332321}, {4.73, 0.0330831}, {4.74, 0.0329348}, {4.75, 0.0327873}, {4.76, \
0.0326404}, {4.77, 0.0324943}, {4.78, 0.0323489}, {4.79, 0.0322042}, {4.8, \
0.0320602}, {4.81, 0.0319169}, {4.82, 0.0317743}, {4.83, 0.0316324}, {4.84, \
0.0314912}, {4.85, 0.0313507}, {4.86, 0.0312108}, {4.87, 0.0310717}, {4.88, \
0.0309333}, {4.89, 0.0307955}, {4.9, 0.0306584}, {4.91, 0.030522}, {4.92, \
0.0303862}, {4.93, 0.0302511}, {4.94, 0.0301167}, {4.95, 0.029983}, {4.96, \
0.0298499}, {4.97, 0.0297174}, {4.98, 0.0295856}, {4.99, 0.0294545}, {5.0, \
0.029324}

Profile of the Lyman-beta line for the observation parallel to the field

{0, 0.188063}, {0.01, 0.188061}, {0.02, 0.188055}, {0.03, 0.188044}, {0.04, \
0.18803}, {0.05, 0.188011}, {0.06, 0.187988}, {0.07, 0.187961}, {0.08, \
0.18793}, {0.09, 0.187894}, {0.1, 0.187854}, {0.11, 0.187811}, {0.12, \
0.187763}, {0.13, 0.18771}, {0.14, 0.187654}, {0.15, 0.187594}, {0.16, \
0.187529}, {0.17, 0.18746}, {0.18, 0.187387}, {0.19, 0.18731}, {0.2, \
0.187229}, {0.21, 0.187144}, {0.22, 0.187055}, {0.23, 0.186961}, {0.24, \
0.186863}, {0.25, 0.186762}, {0.26, 0.186656}, {0.27, 0.186546}, {0.28, \
0.186432}, {0.29, 0.186314}, {0.3, 0.186192}, {0.31, 0.186066}, {0.32, \
0.185936}, {0.33, 0.185801}, {0.34, 0.185663}, {0.35, 0.185521}, {0.36, \
0.185374}, {0.37, 0.185224}, {0.38, 0.18507}, {0.39, 0.184912}, {0.4, \
0.184749}, {0.41, 0.184583}, {0.42, 0.184413}, {0.43, 0.184239}, {0.44, \
0.184061}, {0.45, 0.183879}, {0.46, 0.183693}, {0.47, 0.183503}, {0.48, \
0.18331}, {0.49, 0.183112}, {0.5, 0.182911}, {0.51, 0.182706}, {0.52, \
0.182497}, {0.53, 0.182284}, {0.54, 0.182068}, {0.55, 0.181847}, {0.56, \

0.181623}, {0.57, 0.181395}, {0.58, 0.181164}, {0.59, 0.180928}, {0.6, \
0.180689}, {0.61, 0.180446}, {0.62, 0.1802}, {0.63, 0.17995}, {0.64, \
0.179696}, {0.65, 0.179439}, {0.66, 0.179178}, {0.67, 0.178913}, {0.68, \
0.178645}, {0.69, 0.178373}, {0.7, 0.178098}, {0.71, 0.177819}, {0.72, \
0.177537}, {0.73, 0.177251}, {0.74, 0.176962}, {0.75, 0.176669}, {0.76, \
0.176373}, {0.77, 0.176073}, {0.78, 0.17577}, {0.79, 0.175464}, {0.8, \
0.175154}, {0.81, 0.174841}, {0.82, 0.174525}, {0.83, 0.174205}, {0.84, \
0.173882}, {0.85, 0.173556}, {0.86, 0.173227}, {0.87, 0.172894}, {0.88, \
0.172558}, {0.89, 0.172219}, {0.9, 0.171877}, {0.91, 0.171532}, {0.92, \
0.171183}, {0.93, 0.170832}, {0.94, 0.170477}, {0.95, 0.170119}, {0.96, \
0.169759}, {0.97, 0.169395}, {0.98, 0.169028}, {0.99, 0.168659}, {1.0, \
0.168286}, {1.01, 0.167911}, {1.02, 0.167533}, {1.03, 0.167151}, {1.04, \
0.166767}, {1.05, 0.166381}, {1.06, 0.165991}, {1.07, 0.165599}, {1.08, \
0.165204}, {1.09, 0.164806}, {1.1, 0.164405}, {1.11, 0.164002}, {1.12, \
0.163596}, {1.13, 0.163188}, {1.14, 0.162776}, {1.15, 0.162363}, {1.16, \
0.161947}, {1.17, 0.161528}, {1.18, 0.161107}, {1.19, 0.160683}, {1.2, \
0.160257}, {1.21, 0.159828}, {1.22, 0.159397}, {1.23, 0.158964}, {1.24, \
0.158528}, {1.25, 0.15809}, {1.26, 0.15765}, {1.27, 0.157208}, {1.28, \
0.156763}, {1.29, 0.156316}, {1.3, 0.155867}, {1.31, 0.155415}, {1.32, \
0.154962}, {1.33, 0.154506}, {1.34, 0.154048}, {1.35, 0.153589}, {1.36, \
0.153127}, {1.37, 0.152663}, {1.38, 0.152197}, {1.39, 0.15173}, {1.4, \
0.15126}, {1.41, 0.150788}, {1.42, 0.150315}, {1.43, 0.14984}, {1.44, \
0.149363}, {1.45, 0.148884}, {1.46, 0.148403}, {1.47, 0.147921}, {1.48, \
0.147437}, {1.49, 0.146951}, {1.5, 0.146464}, {1.51, 0.145975}, {1.52, \
0.145484}, {1.53, 0.144992}, {1.54, 0.144498}, {1.55, 0.144003}, {1.56, \
0.143506}, {1.57, 0.143008}, {1.58, 0.142508}, {1.59, 0.142007}, {1.6, \
0.141505}, {1.61, 0.141001}, {1.62, 0.140496}, {1.63, 0.139989}, {1.64, \
0.139482}, {1.65, 0.138973}, {1.66, 0.138463}, {1.67, 0.137951}, {1.68, \
0.137439}, {1.69, 0.136925}, {1.7, 0.13641}, {1.71, 0.135894}, {1.72, \
0.135377}, {1.73, 0.13486}, {1.74, 0.134341}, {1.75, 0.133821}, {1.76, \
0.1333}, {1.77, 0.132778}, {1.78, 0.132255}, {1.79, 0.131732}, {1.8, \
0.131207}, {1.81, 0.130682}, {1.82, 0.130156}, {1.83, 0.129629}, {1.84, \
0.129102}, {1.85, 0.128573}, {1.86, 0.128045}, {1.87, 0.127515}, {1.88, \
0.126985}, {1.89, 0.126454}, {1.9, 0.125923}, {1.91, 0.125391}, {1.92, \
0.124858}, {1.93, 0.124325}, {1.94, 0.123792}, {1.95, 0.123258}, {1.96, \
0.122723}, {1.97, 0.122189}, {1.98, 0.121654}, {1.99, 0.121118}, {2.0, \
0.120582}, {2.01, 0.120046}, {2.02, 0.11951}, {2.03, 0.118973}, {2.04, \
0.118437}, {2.05, 0.1179}, {2.06, 0.117362}, {2.07, 0.116825}, {2.08, \
0.116288}, {2.09, 0.11575}, {2.1, 0.115212}, {2.11, 0.114675}, {2.12, \
0.114137}, {2.13, 0.113599}, {2.14, 0.113062}, {2.15, 0.112524}, {2.16, \
0.111986}, {2.17, 0.111449}, {2.18, 0.110912}, {2.19, 0.110374}, {2.2, \
0.109837}, {2.21, 0.1093}, {2.22, 0.108764}, {2.23, 0.108227}, {2.24, \
0.107691}, {2.25, 0.107155}, {2.26, 0.10662}, {2.27, 0.106084}, {2.28, \
0.105549}, {2.29, 0.105015}, {2.3, 0.10448}, {2.31, 0.103947}, {2.32, \
0.103413}, {2.33, 0.10288}, {2.34, 0.102348}, {2.35, 0.101816}, {2.36, \

0.101285}, {2.37, 0.100754}, {2.38, 0.100223}, {2.39, 0.0996935}, {2.4, \
0.0991643}, {2.41, 0.0986357}, {2.42, 0.0981078}, {2.43, 0.0975805}, {2.44, \
0.0970539}, {2.45, 0.096528}, {2.46, 0.0960029}, {2.47, 0.0954784}, {2.48, \
0.0949547}, {2.49, 0.0944318}, {2.5, 0.0939097}, {2.51, 0.0933884}, {2.52, \
0.0928679}, {2.53, 0.0923483}, {2.54, 0.0918295}, {2.55, 0.0913116}, {2.56, \
0.0907946}, {2.57, 0.0902786}, {2.58, 0.0897635}, {2.59, 0.0892493}, {2.6, \
0.0887361}, {2.61, 0.0882239}, {2.62, 0.0877127}, {2.63, 0.0872026}, {2.64, \
0.0866934}, {2.65, 0.0861854}, {2.66, 0.0856784}, {2.67, 0.0851725}, {2.68, \
0.0846677}, {2.69, 0.084164}, {2.7, 0.0836614}, {2.71, 0.08316}, {2.72, \
0.0826598}, {2.73, 0.0821608}, {2.74, 0.0816629}, {2.75, 0.0811663}, {2.76, \
0.0806709}, {2.77, 0.0801768}, {2.78, 0.0796839}, {2.79, 0.0791922}, {2.8, \
0.0787019}, {2.81, 0.0782128}, {2.82, 0.0777251}, {2.83, 0.0772387}, {2.84, \
0.0767536}, {2.85, 0.0762699}, {2.86, 0.0757875}, {2.87, 0.0753065}, {2.88, \
0.0748269}, {2.89, 0.0743488}, {2.9, 0.073872}, {2.91, 0.0733966}, {2.92, \
0.0729227}, {2.93, 0.0724503}, {2.94, 0.0719793}, {2.95, 0.0715097}, {2.96, \
0.0710417}, {2.97, 0.0705751}, {2.98, 0.0701101}, {2.99, 0.0696466}, {3.0, \
0.0691846}, {3.01, 0.0687241}, {3.02, 0.0682652}, {3.03, 0.0678079}, {3.04, \
0.0673521}, {3.05, 0.0668979}, {3.06, 0.0664452}, {3.07, 0.0659942}, {3.08, \
0.0655448}, {3.09, 0.065097}, {3.1, 0.0646508}, {3.11, 0.0642062}, {3.12, \
0.0637633}, {3.13, 0.063322}, {3.14, 0.0628824}, {3.15, 0.0624445}, {3.16, \
0.0620082}, {3.17, 0.0615736}, {3.18, 0.0611407}, {3.19, 0.0607095}, {3.2, \
0.06028}, {3.21, 0.0598522}, {3.22, 0.0594261}, {3.23, 0.0590017}, {3.24, \
0.0585791}, {3.25, 0.0581582}, {3.26, 0.057739}, {3.27, 0.0573216}, {3.28, \
0.056906}, {3.29, 0.0564921}, {3.3, 0.0560799}, {3.31, 0.0556696}, {3.32, \
0.055261}, {3.33, 0.0548542}, {3.34, 0.0544491}, {3.35, 0.0540459}, {3.36, \
0.0536445}, {3.37, 0.0532448}, {3.38, 0.052847}, {3.39, 0.0524509}, {3.4, \
0.0520567}, {3.41, 0.0516643}, {3.42, 0.0512737}, {3.43, 0.0508849}, {3.44, \
0.050498}, {3.45, 0.0501129}, {3.46, 0.0497296}, {3.47, 0.0493481}, {3.48, \
0.0489685}, {3.49, 0.0485908}, {3.5, 0.0482148}, {3.51, 0.0478408}, {3.52, \
0.0474685}, {3.53, 0.0470981}, {3.54, 0.0467296}, {3.55, 0.0463629}, {3.56, \
0.0459981}, {3.57, 0.0456351}, {3.58, 0.045274}, {3.59, 0.0449148}, {3.6, \
0.0445574}, {3.61, 0.0442019}, {3.62, 0.0438482}, {3.63, 0.0434964}, {3.64, \
0.0431465}, {3.65, 0.0427984}, {3.66, 0.0424522}, {3.67, 0.0421078}, {3.68, \
0.0417653}, {3.69, 0.0414247}, {3.7, 0.041086}, {3.71, 0.0407491}, {3.72, \
0.0404141}, {3.73, 0.0400809}, {3.74, 0.0397496}, {3.75, 0.0394202}, {3.76, \
0.0390926}, {3.77, 0.0387669}, {3.78, 0.0384431}, {3.79, 0.0381211}, {3.8, \
0.0378009}, {3.81, 0.0374826}, {3.82, 0.0371662}, {3.83, 0.0368516}, {3.84, \
0.0365389}, {3.85, 0.036228}, {3.86, 0.035919}, {3.87, 0.0356118}, {3.88, \
0.0353065}, {3.89, 0.035003}, {3.9, 0.0347013}, {3.91, 0.0344015}, {3.92, \
0.0341035}, {3.93, 0.0338073}, {3.94, 0.033513}, {3.95, 0.0332205}, {3.96, \
0.0329298}, {3.97, 0.0326409}, {3.98, 0.0323539}, {3.99, 0.0320686}, {4.0, \
0.0317852}, {4.01, 0.0315036}, {4.02, 0.0312237}, {4.03, 0.0309457}, {4.04, \
0.0306694}, {4.05, 0.030395}, {4.06, 0.0301223}, {4.07, 0.0298515}, {4.08, \
0.0295824}, {4.09, 0.029315}, {4.1, 0.0290495}, {4.11, 0.0287857}, {4.12, \
0.0285237}, {4.13, 0.0282634}, {4.14, 0.0280049}, {4.15, 0.0277481}, {4.16, \

0.0274931}, {4.17, 0.0272398}, {4.18, 0.0269882}, {4.19, 0.0267384}, {4.2, \
0.0264903}, {4.21, 0.0262439}, {4.22, 0.0259992}, {4.23, 0.0257563}, {4.24, \
0.025515}, {4.25, 0.0252755}, {4.26, 0.0250376}, {4.27, 0.0248014}, {4.28, \
0.0245669}, {4.29, 0.0243341}, {4.3, 0.0241029}, {4.31, 0.0238734}, {4.32, \
0.0236456}, {4.33, 0.0234195}, {4.34, 0.0231949}, {4.35, 0.022972}, {4.36, \
0.0227508}, {4.37, 0.0225312}, {4.38, 0.0223132}, {4.39, 0.0220968}, {4.4, \
0.0218821}, {4.41, 0.0216689}, {4.42, 0.0214573}, {4.43, 0.0212474}, {4.44, \
0.021039}, {4.45, 0.0208322}, {4.46, 0.020627}, {4.47, 0.0204233}, {4.48, \
0.0202212}, {4.49, 0.0200207}, {4.5, 0.0198217}, {4.51, 0.0196243}, {4.52, \
0.0194284}, {4.53, 0.019234}, {4.54, 0.0190411}, {4.55, 0.0188498}, {4.56, \
0.0186599}, {4.57, 0.0184716}, {4.58, 0.0182847}, {4.59, 0.0180994}, {4.6, \
0.0179155}, {4.61, 0.0177331}, {4.62, 0.0175522}, {4.63, 0.0173727}, {4.64, \
0.0171947}, {4.65, 0.0170181}, {4.66, 0.016843}, {4.67, 0.0166693}, {4.68, \
0.016497}, {4.69, 0.0163261}, {4.7, 0.0161567}, {4.71, 0.0159886}, {4.72, \
0.015822}, {4.73, 0.0156567}, {4.74, 0.0154928}, {4.75, 0.0153303}, {4.76, \
0.0151692}, {4.77, 0.0150094}, {4.78, 0.014851}, {4.79, 0.0146939}, {4.8, \
0.0145382}, {4.81, 0.0143838}, {4.82, 0.0142307}, {4.83, 0.0140789}, {4.84, \
0.0139285}, {4.85, 0.0137793}, {4.86, 0.0136314}, {4.87, 0.0134849}, {4.88, \
0.0133396}, {4.89, 0.0131955}, {4.9, 0.0130528}, {4.91, 0.0129113}, {4.92, \
0.012771}, {4.93, 0.012632}, {4.94, 0.0124942}, {4.95, 0.0123577}, {4.96, \
0.0122224}, {4.97, 0.0120882}, {4.98, 0.0119553}, {4.99, 0.0118236}, {5.0, \
0.0116931}

Profile of the Lyman-delta line for the observation perpendicular to the field

{0, 0.0582996}, {0.01, 0.0582994}, {0.02, 0.058299}, {0.03, 0.0582983}, \
{0.04, 0.0582972}, {0.05, 0.0582959}, {0.06, 0.0582943}, {0.07, 0.0582924}, \
{0.08, 0.0582902}, {0.09, 0.0582877}, {0.1, 0.0582849}, {0.11, 0.0582818}, \
{0.12, 0.0582784}, {0.13, 0.0582747}, {0.14, 0.0582707}, {0.15, 0.0582665}, \
{0.16, 0.0582619}, {0.17, 0.0582571}, {0.18, 0.0582519}, {0.19, 0.0582465}, \
{0.2, 0.0582407}, {0.21, 0.0582347}, {0.22, 0.0582284}, {0.23, 0.0582218}, \
{0.24, 0.0582149}, {0.25, 0.0582077}, {0.26, 0.0582002}, {0.27, 0.0581924}, \
{0.28, 0.0581843}, {0.29, 0.058176}, {0.3, 0.0581673}, {0.31, 0.0581584}, \
{0.32, 0.0581491}, {0.33, 0.0581396}, {0.34, 0.0581298}, {0.35, 0.0581197}, \
{0.36, 0.0581093}, {0.37, 0.0580986}, {0.38, 0.0580876}, {0.39, 0.0580763}, \
{0.4, 0.0580648}, {0.41, 0.0580529}, {0.42, 0.0580408}, {0.43, 0.0580283}, \
{0.44, 0.0580156}, {0.45, 0.0580026}, {0.46, 0.0579893}, {0.47, 0.0579757}, \
{0.48, 0.0579619}, {0.49, 0.0579477}, {0.5, 0.0579333}, {0.51, 0.0579185}, \
{0.52, 0.0579035}, {0.53, 0.0578882}, {0.54, 0.0578726}, {0.55, 0.0578568}, \
{0.56, 0.0578406}, {0.57, 0.0578242}, {0.58, 0.0578075}, {0.59, 0.0577905}, \
{0.6, 0.0577732}, {0.61, 0.0577556}, {0.62, 0.0577377}, {0.63, 0.0577196}, \
{0.64, 0.0577012}, {0.65, 0.0576825}, {0.66, 0.0576635}, {0.67, 0.0576443}, \
{0.68, 0.0576247}, {0.69, 0.0576049}, {0.7, 0.0575848}, {0.71, 0.0575644}, \
{0.72, 0.0575438}, {0.73, 0.0575228}, {0.74, 0.0575016}, {0.75, 0.0574802}, \
{0.76, 0.0574584}, {0.77, 0.0574364}, {0.78, 0.057414}, {0.79, 0.0573915}, \
{0.8, 0.0573686}, {0.81, 0.0573455}, {0.82, 0.0573221}, {0.83, 0.0572984}, \
{0.84, 0.0572744}, {0.85, 0.0572502}, {0.86, 0.0572257}, {0.87, 0.0572009}, \

```
{0.88, 0.0571759}, {0.89, 0.0571506}, {0.9, 0.057125}, {0.91, 0.0570992}, \
{0.92, 0.0570731}, {0.93, 0.0570467}, {0.94, 0.05702}, {0.95, 0.0569931}, \
{0.96, 0.0569659}, {0.97, 0.0569385}, {0.98, 0.0569108}, {0.99, 0.0568828}, \
{1.0, 0.0568546}, {1.01, 0.0568261}, {1.02, 0.0567973}, {1.03, 0.0567683}, \
{1.04, 0.056739}, {1.05, 0.0567095}, {1.06, 0.0566797}, {1.07, 0.0566496}, \
{1.08, 0.0566193}, {1.09, 0.0565887}, {1.1, 0.0565579}, {1.11, 0.0565268}, \
{1.12, 0.0564955}, {1.13, 0.0564639}, {1.14, 0.056432}, {1.15, 0.0563999}, \
{1.16, 0.0563676}, {1.17, 0.056335}, {1.18, 0.0563021}, {1.19, 0.056269}, \
{1.2, 0.0562356}, {1.21, 0.056202}, {1.22, 0.0561682}, {1.23, 0.0561341}, \
{1.24, 0.0560997}, {1.25, 0.0560651}, {1.26, 0.0560303}, {1.27, 0.0559952}, \
{1.28, 0.0559599}, {1.29, 0.0559243}, {1.3, 0.0558885}, {1.31, 0.0558524}, \
{1.32, 0.0558161}, {1.33, 0.0557796}, {1.34, 0.0557428}, {1.35, 0.0557058}, \
{1.36, 0.0556685}, {1.37, 0.055631}, {1.38, 0.0555933}, {1.39, 0.0555553}, \
{1.4, 0.0555171}, {1.41, 0.0554787}, {1.42, 0.05544}, {1.43, 0.0554011}, \
{1.44, 0.055362}, {1.45, 0.0553226}, {1.46, 0.055283}, {1.47, 0.0552432}, \
{1.48, 0.0552032}, {1.49, 0.0551629}, {1.5, 0.0551224}, {1.51, 0.0550816}, \
{1.52, 0.0550407}, {1.53, 0.0549995}, {1.54, 0.0549581}, {1.55, 0.0549165}, \
{1.56, 0.0548747}, {1.57, 0.0548326}, {1.58, 0.0547903}, {1.59, 0.0547478}, \
{1.6, 0.0547051}, {1.61, 0.0546622}, {1.62, 0.054619}, {1.63, 0.0545756}, \
{1.64, 0.0545321}, {1.65, 0.0544883}, {1.66, 0.0544443}, {1.67, 0.0544}, \
{1.68, 0.0543556}, {1.69, 0.054311}, {1.7, 0.0542661}, {1.71, 0.0542211}, \
{1.72, 0.0541758}, {1.73, 0.0541304}, {1.74, 0.0540847}, {1.75, 0.0540388}, \
{1.76, 0.0539927}, {1.77, 0.0539464}, {1.78, 0.0539}, {1.79, 0.0538533}, \
{1.8, 0.0538064}, {1.81, 0.0537593}, {1.82, 0.053712}, {1.83, 0.0536646}, \
{1.84, 0.0536169}, {1.85, 0.053569}, {1.86, 0.053521}, {1.87, 0.0534727}, \
{1.88, 0.0534243}, {1.89, 0.0533757}, {1.9, 0.0533268}, {1.91, 0.0532778}, \
{1.92, 0.0532286}, {1.93, 0.0531792}, {1.94, 0.0531297}, {1.95, 0.0530799}, \
{1.96, 0.05303}, {1.97, 0.0529799}, {1.98, 0.0529296}, {1.99, 0.0528791}, \
{2.0, 0.0528284}, {2.01, 0.0527776}, {2.02, 0.0527265}, {2.03, 0.0526753}, \
{2.04, 0.052624}, {2.05, 0.0525724}, {2.06, 0.0525207}, {2.07, 0.0524688}, \
{2.08, 0.0524167}, {2.09, 0.0523645}, {2.1, 0.0523121}, {2.11, 0.0522595}, \
{2.12, 0.0522068}, {2.13, 0.0521539}, {2.14, 0.0521008}, {2.15, 0.0520476}, \
{2.16, 0.0519942}, {2.17, 0.0519406}, {2.18, 0.0518869}, {2.19, 0.051833}, \
{2.2, 0.051779}, {2.21, 0.0517248}, {2.22, 0.0516704}, {2.23, 0.0516159}, \
{2.24, 0.0515612}, {2.25, 0.0515064}, {2.26, 0.0514514}, {2.27, 0.0513963}, \
{2.28, 0.051341}, {2.29, 0.0512856}, {2.3, 0.05123}, {2.31, 0.0511743}, \
{2.32, 0.0511184}, {2.33, 0.0510624}, {2.34, 0.0510062}, {2.35, 0.0509499}, \
{2.36, 0.0508935}, {2.37, 0.0508369}, {2.38, 0.0507801}, {2.39, 0.0507233}, \
{2.4, 0.0506663}, {2.41, 0.0506091}, {2.42, 0.0505518}, {2.43, 0.0504944}, \
{2.44, 0.0504369}, {2.45, 0.0503792}, {2.46, 0.0503214}, {2.47, 0.0502634}, \
{2.48, 0.0502054}, {2.49, 0.0501472}, {2.5, 0.0500888}, {2.51, 0.0500304}, \
{2.52, 0.0499718}, {2.53, 0.0499131}, {2.54, 0.0498543}, {2.55, 0.0497953}, \
{2.56, 0.0497363}, {2.57, 0.0496771}, {2.58, 0.0496178}, {2.59, 0.0495583}, \
{2.6, 0.0494988}, {2.61, 0.0494391}, {2.62, 0.0493794}, {2.63, 0.0493195}, \
{2.64, 0.0492595}, {2.65, 0.0491994}, {2.66, 0.0491392}, {2.67, 0.0490788}, \
```

{2.68, 0.0490184}, {2.69, 0.0489579}, {2.7, 0.0488972}, {2.71, 0.0488364}, \
{2.72, 0.0487756}, {2.73, 0.0487146}, {2.74, 0.0486536}, {2.75, 0.0485924}, \
{2.76, 0.0485312}, {2.77, 0.0484698}, {2.78, 0.0484083}, {2.79, 0.0483468}, \
{2.8, 0.0482851}, {2.81, 0.0482234}, {2.82, 0.0481616}, {2.83, 0.0480996}, \
{2.84, 0.0480376}, {2.85, 0.0479755}, {2.86, 0.0479133}, {2.87, 0.047851}, \
{2.88, 0.0477887}, {2.89, 0.0477262}, {2.9, 0.0476637}, {2.91, 0.0476011}, \
{2.92, 0.0475384}, {2.93, 0.0474756}, {2.94, 0.0474127}, {2.95, 0.0473498}, \
{2.96, 0.0472868}, {2.97, 0.0472237}, {2.98, 0.0471605}, {2.99, 0.0470972}, \
{3.0, 0.0470339}, {3.01, 0.0469705}, {3.02, 0.0469071}, {3.03, 0.0468435}, \
{3.04, 0.0467799}, {3.05, 0.0467162}, {3.06, 0.0466525}, {3.07, 0.0465887}, \
{3.08, 0.0465248}, {3.09, 0.0464609}, {3.1, 0.0463969}, {3.11, 0.0463328}, \
{3.12, 0.0462687}, {3.13, 0.0462045}, {3.14, 0.0461403}, {3.15, 0.046076}, \
{3.16, 0.0460116}, {3.17, 0.0459472}, {3.18, 0.0458828}, {3.19, 0.0458182}, \
{3.2, 0.0457537}, {3.21, 0.045689}, {3.22, 0.0456244}, {3.23, 0.0455596}, \
{3.24, 0.0454949}, {3.25, 0.04543}, {3.26, 0.0453652}, {3.27, 0.0453003}, \
{3.28, 0.0452353}, {3.29, 0.0451703}, {3.3, 0.0451053}, {3.31, 0.0450402}, \
{3.32, 0.0449751}, {3.33, 0.0449099}, {3.34, 0.0448447}, {3.35, 0.0447795}, \
{3.36, 0.0447142}, {3.37, 0.0446489}, {3.38, 0.0445835}, {3.39, 0.0445181}, \
{3.4, 0.0444527}, {3.41, 0.0443873}, {3.42, 0.0443218}, {3.43, 0.0442563}, \
{3.44, 0.0441908}, {3.45, 0.0441252}, {3.46, 0.0440596}, {3.47, 0.043994}, \
{3.48, 0.0439284}, {3.49, 0.0438627}, {3.5, 0.043797}, {3.51, 0.0437313}, \
{3.52, 0.0436656}, {3.53, 0.0435999}, {3.54, 0.0435341}, {3.55, 0.0434683}, \
{3.56, 0.0434025}, {3.57, 0.0433367}, {3.58, 0.0432709}, {3.59, 0.0432051}, \
{3.6, 0.0431392}, {3.61, 0.0430733}, {3.62, 0.0430075}, {3.63, 0.0429416}, \
{3.64, 0.0428757}, {3.65, 0.0428098}, {3.66, 0.0427438}, {3.67, 0.0426779}, \
{3.68, 0.042612}, {3.69, 0.0425461}, {3.7, 0.0424801}, {3.71, 0.0424142}, \
{3.72, 0.0423483}, {3.73, 0.0422823}, {3.74, 0.0422164}, {3.75, 0.0421504}, \
{3.76, 0.0420845}, {3.77, 0.0420186}, {3.78, 0.0419526}, {3.79, 0.0418867}, \
{3.8, 0.0418208}, {3.81, 0.0417549}, {3.82, 0.041689}, {3.83, 0.0416231}, \
{3.84, 0.0415572}, {3.85, 0.0414913}, {3.86, 0.0414254}, {3.87, 0.0413596}, \
{3.88, 0.0412937}, {3.89, 0.0412279}, {3.9, 0.0411621}, {3.91, 0.0410963}, \
{3.92, 0.0410305}, {3.93, 0.0409647}, {3.94, 0.0408989}, {3.95, 0.0408332}, \
{3.96, 0.0407675}, {3.97, 0.0407018}, {3.98, 0.0406361}, {3.99, 0.0405704}, \
{4.0, 0.0405048}, {4.01, 0.0404392}, {4.02, 0.0403736}, {4.03, 0.040308}, \
{4.04, 0.0402425}, {4.05, 0.040177}, {4.06, 0.0401115}, {4.07, 0.040046}, \
{4.08, 0.0399806}, {4.09, 0.0399152}, {4.1, 0.0398498}, {4.11, 0.0397845}, \
{4.12, 0.0397192}, {4.13, 0.0396539}, {4.14, 0.0395887}, {4.15, 0.0395235}, \
{4.16, 0.0394583}, {4.17, 0.0393932}, {4.18, 0.0393281}, {4.19, 0.039263}, \
{4.2, 0.039198}, {4.21, 0.039133}, {4.22, 0.0390681}, {4.23, 0.0390032}, \
{4.24, 0.0389383}, {4.25, 0.0388735}, {4.26, 0.0388087}, {4.27, 0.038744}, \
{4.28, 0.0386793}, {4.29, 0.0386146}, {4.3, 0.03855}, {4.31, 0.0384855}, \
{4.32, 0.038421}, {4.33, 0.0383565}, {4.34, 0.0382921}, {4.35, 0.0382277}, \
{4.36, 0.0381634}, {4.37, 0.0380991}, {4.38, 0.0380349}, {4.39, 0.0379708}, \
{4.4, 0.0379067}, {4.41, 0.0378426}, {4.42, 0.0377786}, {4.43, 0.0377146}, \
{4.44, 0.0376507}, {4.45, 0.0375869}, {4.46, 0.0375231}, {4.47, 0.0374594}, \

{4.48, 0.0373957}, {4.49, 0.0373321}, {4.5, 0.0372686}, {4.51, 0.0372051}, \
{4.52, 0.0371416}, {4.53, 0.0370783}, {4.54, 0.0370149}, {4.55, 0.0369517}, \
{4.56, 0.0368885}, {4.57, 0.0368254}, {4.58, 0.0367623}, {4.59, 0.0366993}, \
{4.6, 0.0366364}, {4.61, 0.0365735}, {4.62, 0.0365107}, {4.63, 0.036448}, \
{4.64, 0.0363853}, {4.65, 0.0363227}, {4.66, 0.0362602}, {4.67, 0.0361977}, \
{4.68, 0.0361354}, {4.69, 0.036073}, {4.7, 0.0360108}, {4.71, 0.0359486}, \
{4.72, 0.0358865}, {4.73, 0.0358245}, {4.74, 0.0357625}, {4.75, 0.0357006}, \
{4.76, 0.0356388}, {4.77, 0.0355771}, {4.78, 0.0355154}, {4.79, 0.0354538}, \
{4.8, 0.0353923}, {4.81, 0.0353309}, {4.82, 0.0352695}, {4.83, 0.0352082}, \
{4.84, 0.035147}, {4.85, 0.0350859}, {4.86, 0.0350248}, {4.87, 0.0349639}, \
{4.88, 0.034903}, {4.89, 0.0348422}, {4.9, 0.0347814}, {4.91, 0.0347208}, \
{4.92, 0.0346602}, {4.93, 0.0345998}, {4.94, 0.0345394}, {4.95, 0.034479}, \
{4.96, 0.0344188}, {4.97, 0.0343587}, {4.98, 0.0342986}, {4.99, 0.0342386}, \
{5.0, 0.0341787}

Profile of the Lyman-delta line for the observation parallel to the field

{0, 0.0827478}, {0.01, 0.0827475}, {0.02, 0.0827467}, {0.03, 0.0827453}, \
{0.04, 0.0827434}, {0.05, 0.0827409}, {0.06, 0.0827378}, {0.07, 0.0827342}, \
{0.08, 0.08273}, {0.09, 0.0827253}, {0.1, 0.0827201}, {0.11, 0.0827142}, \
{0.12, 0.0827079}, {0.13, 0.0827009}, {0.14, 0.0826934}, {0.15, 0.0826854}, \
{0.16, 0.0826768}, {0.17, 0.0826677}, {0.18, 0.082658}, {0.19, 0.0826477}, \
{0.2, 0.0826369}, {0.21, 0.0826255}, {0.22, 0.0826136}, {0.23, 0.0826012}, \
{0.24, 0.0825881}, {0.25, 0.0825746}, {0.26, 0.0825605}, {0.27, 0.0825458}, \
{0.28, 0.0825306}, {0.29, 0.0825148}, {0.3, 0.0824985}, {0.31, 0.0824816}, \
{0.32, 0.0824642}, {0.33, 0.0824463}, {0.34, 0.0824277}, {0.35, 0.0824087}, \
{0.36, 0.0823891}, {0.37, 0.0823689}, {0.38, 0.0823482}, {0.39, 0.082327}, \
{0.4, 0.0823052}, {0.41, 0.0822829}, {0.42, 0.08226}, {0.43, 0.0822366}, \
{0.44, 0.0822126}, {0.45, 0.0821881}, {0.46, 0.082163}, {0.47, 0.0821375}, \
{0.48, 0.0821113}, {0.49, 0.0820847}, {0.5, 0.0820574}, {0.51, 0.0820297}, \
{0.52, 0.0820014}, {0.53, 0.0819726}, {0.54, 0.0819432}, {0.55, 0.0819133}, \
{0.56, 0.0818829}, {0.57, 0.0818519}, {0.58, 0.0818204}, {0.59, 0.0817884}, \
{0.6, 0.0817558}, {0.61, 0.0817227}, {0.62, 0.0816891}, {0.63, 0.0816549}, \
{0.64, 0.0816202}, {0.65, 0.081585}, {0.66, 0.0815493}, {0.67, 0.081513}, \
{0.68, 0.0814762}, {0.69, 0.0814389}, {0.7, 0.081401}, {0.71, 0.0813626}, \
{0.72, 0.0813238}, {0.73, 0.0812843}, {0.74, 0.0812444}, {0.75, 0.0812039}, \
{0.76, 0.081163}, {0.77, 0.0811215}, {0.78, 0.0810795}, {0.79, 0.0810369}, \
{0.8, 0.0809939}, {0.81, 0.0809503}, {0.82, 0.0809062}, {0.83, 0.0808617}, \
{0.84, 0.0808166}, {0.85, 0.080771}, {0.86, 0.0807249}, {0.87, 0.0806782}, \
{0.88, 0.0806311}, {0.89, 0.0805835}, {0.9, 0.0805353}, {0.91, 0.0804867}, \
{0.92, 0.0804375}, {0.93, 0.0803879}, {0.94, 0.0803377}, {0.95, 0.0802871}, \
{0.96, 0.080236}, {0.97, 0.0801843}, {0.98, 0.0801322}, {0.99, 0.0800795}, \
{1.0, 0.0800264}, {1.01, 0.0799728}, {1.02, 0.0799187}, {1.03, 0.0798641}, \
{1.04, 0.079809}, {1.05, 0.0797535}, {1.06, 0.0796974}, {1.07, 0.0796409}, \
{1.08, 0.0795838}, {1.09, 0.0795263}, {1.1, 0.0794683}, {1.11, 0.0794099}, \
{1.12, 0.0793509}, {1.13, 0.0792915}, {1.14, 0.0792316}, {1.15, 0.0791712}, \
{1.16, 0.0791104}, {1.17, 0.0790491}, {1.18, 0.0789873}, {1.19, 0.078925}, \

{1.2, 0.0788623}, {1.21, 0.0787991}, {1.22, 0.0787355}, {1.23, 0.0786714}, \
{1.24, 0.0786068}, {1.25, 0.0785418}, {1.26, 0.0784763}, {1.27, 0.0784103}, \
{1.28, 0.0783439}, {1.29, 0.0782771}, {1.3, 0.0782098}, {1.31, 0.078142}, \
{1.32, 0.0780738}, {1.33, 0.0780052}, {1.34, 0.0779361}, {1.35, 0.0778665}, \
{1.36, 0.0777965}, {1.37, 0.0777261}, {1.38, 0.0776552}, {1.39, 0.0775839}, \
{1.4, 0.0775122}, {1.41, 0.07744}, {1.42, 0.0773674}, {1.43, 0.0772943}, \
{1.44, 0.0772208}, {1.45, 0.0771469}, {1.46, 0.0770726}, {1.47, 0.0769978}, \
{1.48, 0.0769226}, {1.49, 0.076847}, {1.5, 0.076771}, {1.51, 0.0766946}, \
{1.52, 0.0766177}, {1.53, 0.0765404}, {1.54, 0.0764627}, {1.55, 0.0763846}, \
{1.56, 0.0763061}, {1.57, 0.0762272}, {1.58, 0.0761479}, {1.59, 0.0760681}, \
{1.6, 0.075988}, {1.61, 0.0759074}, {1.62, 0.0758265}, {1.63, 0.0757452}, \
{1.64, 0.0756634}, {1.65, 0.0755813}, {1.66, 0.0754988}, {1.67, 0.0754159}, \
{1.68, 0.0753326}, {1.69, 0.0752489}, {1.7, 0.0751648}, {1.71, 0.0750803}, \
{1.72, 0.0749955}, {1.73, 0.0749103}, {1.74, 0.0748247}, {1.75, 0.0747387}, \
{1.76, 0.0746523}, {1.77, 0.0745656}, {1.78, 0.0744785}, {1.79, 0.0743911}, \
{1.8, 0.0743032}, {1.81, 0.074215}, {1.82, 0.0741265}, {1.83, 0.0740375}, \
{1.84, 0.0739483}, {1.85, 0.0738586}, {1.86, 0.0737686}, {1.87, 0.0736783}, \
{1.88, 0.0735876}, {1.89, 0.0734965}, {1.9, 0.0734051}, {1.91, 0.0733134}, \
{1.92, 0.0732213}, {1.93, 0.0731289}, {1.94, 0.0730361}, {1.95, 0.072943}, \
{1.96, 0.0728495}, {1.97, 0.0727558}, {1.98, 0.0726616}, {1.99, 0.0725672}, \
{2.0, 0.0724724}, {2.01, 0.0723773}, {2.02, 0.0722819}, {2.03, 0.0721862}, \
{2.04, 0.0720901}, {2.05, 0.0719937}, {2.06, 0.071897}, {2.07, 0.0718}, \
{2.08, 0.0717027}, {2.09, 0.071605}, {2.1, 0.0715071}, {2.11, 0.0714088}, \
{2.12, 0.0713103}, {2.13, 0.0712114}, {2.14, 0.0711122}, {2.15, 0.0710128}, \
{2.16, 0.070913}, {2.17, 0.070813}, {2.18, 0.0707126}, {2.19, 0.070612}, \
{2.2, 0.0705111}, {2.21, 0.0704099}, {2.22, 0.0703084}, {2.23, 0.0702066}, \
{2.24, 0.0701046}, {2.25, 0.0700022}, {2.26, 0.0698996}, {2.27, 0.0697968}, \
{2.28, 0.0696936}, {2.29, 0.0695902}, {2.3, 0.0694865}, {2.31, 0.0693825}, \
{2.32, 0.0692783}, {2.33, 0.0691739}, {2.34, 0.0690691}, {2.35, 0.0689641}, \
{2.36, 0.0688589}, {2.37, 0.0687534}, {2.38, 0.0686476}, {2.39, 0.0685416}, \
{2.4, 0.0684354}, {2.41, 0.0683289}, {2.42, 0.0682222}, {2.43, 0.0681152}, \
{2.44, 0.068008}, {2.45, 0.0679005}, {2.46, 0.0677929}, {2.47, 0.0676849}, \
{2.48, 0.0675768}, {2.49, 0.0674684}, {2.5, 0.0673598}, {2.51, 0.067251}, \
{2.52, 0.067142}, {2.53, 0.0670327}, {2.54, 0.0669233}, {2.55, 0.0668136}, \
{2.56, 0.0667037}, {2.57, 0.0665936}, {2.58, 0.0664832}, {2.59, 0.0663727}, \
{2.6, 0.066262}, {2.61, 0.066151}, {2.62, 0.0660399}, {2.63, 0.0659286}, \
{2.64, 0.0658171}, {2.65, 0.0657053}, {2.66, 0.0655934}, {2.67, 0.0654813}, \
{2.68, 0.065369}, {2.69, 0.0652566}, {2.7, 0.0651439}, {2.71, 0.0650311}, \
{2.72, 0.0649181}, {2.73, 0.0648049}, {2.74, 0.0646915}, {2.75, 0.064578}, \
{2.76, 0.0644643}, {2.77, 0.0643504}, {2.78, 0.0642364}, {2.79, 0.0641221}, \
{2.8, 0.0640078}, {2.81, 0.0638933}, {2.82, 0.0637786}, {2.83, 0.0636637}, \
{2.84, 0.0635487}, {2.85, 0.0634336}, {2.86, 0.0633183}, {2.87, 0.0632029}, \
{2.88, 0.0630873}, {2.89, 0.0629716}, {2.9, 0.0628557}, {2.91, 0.0627397}, \
{2.92, 0.0626235}, {2.93, 0.0625073}, {2.94, 0.0623909}, {2.95, 0.0622743}, \
{2.96, 0.0621576}, {2.97, 0.0620409}, {2.98, 0.0619239}, {2.99, 0.0618069}, \

{3.0, 0.0616897}, {3.01, 0.0615725}, {3.02, 0.0614551}, {3.03, 0.0613376}, \
{3.04, 0.0612199}, {3.05, 0.0611022}, {3.06, 0.0609844}, {3.07, 0.0608664}, \
{3.08, 0.0607484}, {3.09, 0.0606303}, {3.1, 0.060512}, {3.11, 0.0603937}, \
{3.12, 0.0602753}, {3.13, 0.0601567}, {3.14, 0.0600381}, {3.15, 0.0599194}, \
{3.16, 0.0598006}, {3.17, 0.0596817}, {3.18, 0.0595628}, {3.19, 0.0594437}, \
{3.2, 0.0593246}, {3.21, 0.0592054}, {3.22, 0.0590861}, {3.23, 0.0589668}, \
{3.24, 0.0588474}, {3.25, 0.0587279}, {3.26, 0.0586084}, {3.27, 0.0584887}, \
{3.28, 0.0583691}, {3.29, 0.0582493}, {3.3, 0.0581295}, {3.31, 0.0580097}, \
{3.32, 0.0578898}, {3.33, 0.0577698}, {3.34, 0.0576498}, {3.35, 0.0575297}, \
{3.36, 0.0574096}, {3.37, 0.0572895}, {3.38, 0.0571693}, {3.39, 0.0570491}, \
{3.4, 0.0569288}, {3.41, 0.0568085}, {3.42, 0.0566881}, {3.43, 0.0565677}, \
{3.44, 0.0564473}, {3.45, 0.0563269}, {3.46, 0.0562064}, {3.47, 0.0560859}, \
{3.48, 0.0559654}, {3.49, 0.0558449}, {3.5, 0.0557243}, {3.51, 0.0556038}, \
{3.52, 0.0554832}, {3.53, 0.0553626}, {3.54, 0.0552419}, {3.55, 0.0551213}, \
{3.56, 0.0550007}, {3.57, 0.05488}, {3.58, 0.0547594}, {3.59, 0.0546387}, \
{3.6, 0.0545181}, {3.61, 0.0543974}, {3.62, 0.0542768}, {3.63, 0.0541561}, \
{3.64, 0.0540355}, {3.65, 0.0539149}, {3.66, 0.0537942}, {3.67, 0.0536736}, \
{3.68, 0.053553}, {3.69, 0.0534325}, {3.7, 0.0533119}, {3.71, 0.0531913}, \
{3.72, 0.0530708}, {3.73, 0.0529503}, {3.74, 0.0528298}, {3.75, 0.0527094}, \
{3.76, 0.052589}, {3.77, 0.0524686}, {3.78, 0.0523482}, {3.79, 0.0522279}, \
{3.8, 0.0521076}, {3.81, 0.0519873}, {3.82, 0.0518671}, {3.83, 0.0517469}, \
{3.84, 0.0516268}, {3.85, 0.0515067}, {3.86, 0.0513866}, {3.87, 0.0512666}, \
{3.88, 0.0511466}, {3.89, 0.0510267}, {3.9, 0.0509069}, {3.91, 0.0507871}, \
{3.92, 0.0506673}, {3.93, 0.0505476}, {3.94, 0.050428}, {3.95, 0.0503084}, \
{3.96, 0.0501889}, {3.97, 0.0500694}, {3.98, 0.0499501}, {3.99, 0.0498307}, \
{4.0, 0.0497115}, {4.01, 0.0495923}, {4.02, 0.0494732}, {4.03, 0.0493541}, \
{4.04, 0.0492352}, {4.05, 0.0491163}, {4.06, 0.0489975}, {4.07, 0.0488787}, \
{4.08, 0.0487601}, {4.09, 0.0486415}, {4.1, 0.048523}, {4.11, 0.0484046}, \
{4.12, 0.0482863}, {4.13, 0.0481681}, {4.14, 0.0480499}, {4.15, 0.0479319}, \
{4.16, 0.0478139}, {4.17, 0.047696}, {4.18, 0.0475783}, {4.19, 0.0474606}, \
{4.2, 0.047343}, {4.21, 0.0472255}, {4.22, 0.0471082}, {4.23, 0.0469909}, \
{4.24, 0.0468737}, {4.25, 0.0467567}, {4.26, 0.0466397}, {4.27, 0.0465229}, \
{4.28, 0.0464061}, {4.29, 0.0462895}, {4.3, 0.046173}, {4.31, 0.0460566}, \
{4.32, 0.0459403}, {4.33, 0.0458241}, {4.34, 0.0457081}, {4.35, 0.0455921}, \
{4.36, 0.0454763}, {4.37, 0.0453606}, {4.38, 0.045245}, {4.39, 0.0451296}, \
{4.4, 0.0450143}, {4.41, 0.0448991}, {4.42, 0.044784}, {4.43, 0.0446691}, \
{4.44, 0.0445543}, {4.45, 0.0444396}, {4.46, 0.0443251}, {4.47, 0.0442107}, \
{4.48, 0.0440964}, {4.49, 0.0439823}, {4.5, 0.0438683}, {4.51, 0.0437544}, \
{4.52, 0.0436407}, {4.53, 0.0435271}, {4.54, 0.0434137}, {4.55, 0.0433004}, \
{4.56, 0.0431873}, {4.57, 0.0430743}, {4.58, 0.0429614}, {4.59, 0.0428487}, \
{4.6, 0.0427362}, {4.61, 0.0426238}, {4.62, 0.0425115}, {4.63, 0.0423994}, \
{4.64, 0.0422875}, {4.65, 0.0421757}, {4.66, 0.0420641}, {4.67, 0.0419526}, \
{4.68, 0.0418413}, {4.69, 0.0417302}, {4.7, 0.0416192}, {4.71, 0.0415084}, \
{4.72, 0.0413977}, {4.73, 0.0412872}, {4.74, 0.0411769}, {4.75, 0.0410667}, \
{4.76, 0.0409567}, {4.77, 0.0408469}, {4.78, 0.0407372}, {4.79, 0.0406277}, \

{4.8, 0.0405184}, {4.81, 0.0404092}, {4.82, 0.0403003}, {4.83, 0.0401915}, \
{4.84, 0.0400828}, {4.85, 0.0399744}, {4.86, 0.0398661}, {4.87, 0.039758}, \
{4.88, 0.0396501}, {4.89, 0.0395424}, {4.9, 0.0394348}, {4.91, 0.0393274}, \
{4.92, 0.0392202}, {4.93, 0.0391132}, {4.94, 0.0390064}, {4.95, 0.0388998}, \
{4.96, 0.0387933}, {4.97, 0.038687}, {4.98, 0.038581}, {4.99, 0.0384751}, \
{5.0, 0.0383694}

Profile of the Balmer-beta line for the observation perpendicular to the field

{0, 0.101709}, {0.01, 0.101709}, {0.02, 0.101707}, {0.03, 0.101703}, {0.04, \
0.101699}, {0.05, 0.101693}, {0.06, 0.101686}, {0.07, 0.101677}, {0.08, \
0.101668}, {0.09, 0.101656}, {0.1, 0.101644}, {0.11, 0.10163}, {0.12, \
0.101615}, {0.13, 0.101599}, {0.14, 0.101582}, {0.15, 0.101563}, {0.16, \
0.101543}, {0.17, 0.101521}, {0.18, 0.101499}, {0.19, 0.101474}, {0.2, \
0.101449}, {0.21, 0.101423}, {0.22, 0.101395}, {0.23, 0.101366}, {0.24, \
0.101335}, {0.25, 0.101304}, {0.26, 0.101271}, {0.27, 0.101236}, {0.28, \
0.101201}, {0.29, 0.101164}, {0.3, 0.101126}, {0.31, 0.101087}, {0.32, \
0.101046}, {0.33, 0.101005}, {0.34, 0.100962}, {0.35, 0.100918}, {0.36, \
0.100872}, {0.37, 0.100825}, {0.38, 0.100777}, {0.39, 0.100728}, {0.4, \
0.100678}, {0.41, 0.100626}, {0.42, 0.100574}, {0.43, 0.10052}, {0.44, \
0.100464}, {0.45, 0.100408}, {0.46, 0.10035}, {0.47, 0.100292}, {0.48, \
0.100232}, {0.49, 0.100171}, {0.5, 0.100108}, {0.51, 0.100045}, {0.52, \
0.0999801}, {0.53, 0.0999143}, {0.54, 0.0998474}, {0.55, 0.0997793}, {0.56, \
0.0997101}, {0.57, 0.0996397}, {0.58, 0.0995682}, {0.59, 0.0994956}, {0.6, \
0.0994218}, {0.61, 0.099347}, {0.62, 0.099271}, {0.63, 0.0991939}, {0.64, \
0.0991158}, {0.65, 0.0990366}, {0.66, 0.0989562}, {0.67, 0.0988748}, {0.68, \
0.0987924}, {0.69, 0.0987089}, {0.7, 0.0986243}, {0.71, 0.0985387}, {0.72, \
0.098452}, {0.73, 0.0983643}, {0.74, 0.0982756}, {0.75, 0.0981858}, {0.76, \
0.098095}, {0.77, 0.0980033}, {0.78, 0.0979105}, {0.79, 0.0978167}, {0.8, \
0.097722}, {0.81, 0.0976263}, {0.82, 0.0975296}, {0.83, 0.0974319}, {0.84, \
0.0973333}, {0.85, 0.0972337}, {0.86, 0.0971332}, {0.87, 0.0970318}, {0.88, \
0.0969294}, {0.89, 0.0968261}, {0.9, 0.0967219}, {0.91, 0.0966168}, {0.92, \
0.0965108}, {0.93, 0.096404}, {0.94, 0.0962962}, {0.95, 0.0961875}, {0.96, \
0.096078}, {0.97, 0.0959677}, {0.98, 0.0958565}, {0.99, 0.0957444}, {1.0, \
0.0956315}, {1.01, 0.0955178}, {1.02, 0.0954033}, {1.03, 0.095288}, {1.04, \
0.0951718}, {1.05, 0.0950549}, {1.06, 0.0949372}, {1.07, 0.0948187}, {1.08, \
0.0946994}, {1.09, 0.0945794}, {1.1, 0.0944586}, {1.11, 0.0943371}, {1.12, \
0.0942148}, {1.13, 0.0940918}, {1.14, 0.0939681}, {1.15, 0.0938437}, {1.16, \
0.0937186}, {1.17, 0.0935928}, {1.18, 0.0934662}, {1.19, 0.093339}, {1.2, \
0.0932112}, {1.21, 0.0930827}, {1.22, 0.0929535}, {1.23, 0.0928236}, {1.24, \
0.0926932}, {1.25, 0.0925621}, {1.26, 0.0924303}, {1.27, 0.092298}, {1.28, \
0.092165}, {1.29, 0.0920315}, {1.3, 0.0918973}, {1.31, 0.0917626}, {1.32, \
0.0916273}, {1.33, 0.0914914}, {1.34, 0.091355}, {1.35, 0.091218}, {1.36, \
0.0910805}, {1.37, 0.0909424}, {1.38, 0.0908038}, {1.39, 0.0906647}, {1.4, \
0.0905251}, {1.41, 0.090385}, {1.42, 0.0902444}, {1.43, 0.0901033}, {1.44, \
0.0899617}, {1.45, 0.0898196}, {1.46, 0.0896771}, {1.47, 0.0895341}, {1.48, \
0.0893906}, {1.49, 0.0892468}, {1.5, 0.0891024}, {1.51, 0.0889577}, {1.52, \

0.0888125}, {1.53, 0.088667}, {1.54, 0.088521}, {1.55, 0.0883746}, {1.56, \
0.0882278}, {1.57, 0.0880807}, {1.58, 0.0879331}, {1.59, 0.0877852}, {1.6, \
0.087637}, {1.61, 0.0874883}, {1.62, 0.0873394}, {1.63, 0.0871901}, {1.64, \
0.0870404}, {1.65, 0.0868905}, {1.66, 0.0867402}, {1.67, 0.0865896}, {1.68, \
0.0864387}, {1.69, 0.0862875}, {1.7, 0.086136}, {1.71, 0.0859842}, {1.72, \
0.0858321}, {1.73, 0.0856798}, {1.74, 0.0855272}, {1.75, 0.0853743}, {1.76, \
0.0852212}, {1.77, 0.0850679}, {1.78, 0.0849143}, {1.79, 0.0847604}, {1.8, \
0.0846064}, {1.81, 0.0844521}, {1.82, 0.0842976}, {1.83, 0.0841429}, {1.84, \
0.0839879}, {1.85, 0.0838328}, {1.86, 0.0836775}, {1.87, 0.083522}, {1.88, \
0.0833664}, {1.89, 0.0832105}, {1.9, 0.0830545}, {1.91, 0.0828984}, {1.92, \
0.082742}, {1.93, 0.0825855}, {1.94, 0.0824289}, {1.95, 0.0822722}, {1.96, \
0.0821153}, {1.97, 0.0819582}, {1.98, 0.0818011}, {1.99, 0.0816438}, {2.0, \
0.0814864}, {2.01, 0.0813289}, {2.02, 0.0811713}, {2.03, 0.0810136}, {2.04, \
0.0808558}, {2.05, 0.0806979}, {2.06, 0.08054}, {2.07, 0.0803819}, {2.08, \
0.0802238}, {2.09, 0.0800656}, {2.1, 0.0799074}, {2.11, 0.079749}, {2.12, \
0.0795907}, {2.13, 0.0794322}, {2.14, 0.0792738}, {2.15, 0.0791153}, {2.16, \
0.0789567}, {2.17, 0.0787981}, {2.18, 0.0786395}, {2.19, 0.0784809}, {2.2, \
0.0783222}, {2.21, 0.0781635}, {2.22, 0.0780048}, {2.23, 0.0778461}, {2.24, \
0.0776874}, {2.25, 0.0775287}, {2.26, 0.07737}, {2.27, 0.0772113}, {2.28, \
0.0770526}, {2.29, 0.0768939}, {2.3, 0.0767353}, {2.31, 0.0765766}, {2.32, \
0.076418}, {2.33, 0.0762594}, {2.34, 0.0761008}, {2.35, 0.0759423}, {2.36, \
0.0757838}, {2.37, 0.0756254}, {2.38, 0.075467}, {2.39, 0.0753086}, {2.4, \
0.0751503}, {2.41, 0.074992}, {2.42, 0.0748338}, {2.43, 0.0746757}, {2.44, \
0.0745176}, {2.45, 0.0743596}, {2.46, 0.0742016}, {2.47, 0.0740437}, {2.48, \
0.0738859}, {2.49, 0.0737282}, {2.5, 0.0735706}, {2.51, 0.073413}, {2.52, \
0.0732555}, {2.53, 0.0730981}, {2.54, 0.0729408}, {2.55, 0.0727836}, {2.56, \
0.0726264}, {2.57, 0.0724694}, {2.58, 0.0723125}, {2.59, 0.0721556}, {2.6, \
0.0719989}, {2.61, 0.0718423}, {2.62, 0.0716858}, {2.63, 0.0715293}, {2.64, \
0.071373}, {2.65, 0.0712169}, {2.66, 0.0710608}, {2.67, 0.0709048}, {2.68, \
0.070749}, {2.69, 0.0705933}, {2.7, 0.0704377}, {2.71, 0.0702822}, {2.72, \
0.0701269}, {2.73, 0.0699717}, {2.74, 0.0698166}, {2.75, 0.0696617}, {2.76, \
0.0695068}, {2.77, 0.0693522}, {2.78, 0.0691976}, {2.79, 0.0690432}, {2.8, \
0.068889}, {2.81, 0.0687348}, {2.82, 0.0685809}, {2.83, 0.068427}, {2.84, \
0.0682733}, {2.85, 0.0681198}, {2.86, 0.0679664}, {2.87, 0.0678132}, {2.88, \
0.0676601}, {2.89, 0.0675071}, {2.9, 0.0673544}, {2.91, 0.0672017}, {2.92, \
0.0670493}, {2.93, 0.066897}, {2.94, 0.0667448}, {2.95, 0.0665928}, {2.96, \
0.066441}, {2.97, 0.0662893}, {2.98, 0.0661378}, {2.99, 0.0659865}, {3.0, \
0.0658353}, {3.01, 0.0656843}, {3.02, 0.0655334}, {3.03, 0.0653828}, {3.04, \
0.0652323}, {3.05, 0.0650819}, {3.06, 0.0649318}, {3.07, 0.0647818}, {3.08, \
0.064632}, {3.09, 0.0644823}, {3.1, 0.0643329}, {3.11, 0.0641836}, {3.12, \
0.0640345}, {3.13, 0.0638855}, {3.14, 0.0637368}, {3.15, 0.0635882}, {3.16, \
0.0634398}, {3.17, 0.0632916}, {3.18, 0.0631436}, {3.19, 0.0629957}, {3.2, \
0.062848}, {3.21, 0.0627006}, {3.22, 0.0625533}, {3.23, 0.0624061}, {3.24, \
0.0622592}, {3.25, 0.0621125}, {3.26, 0.0619659}, {3.27, 0.0618195}, {3.28, \
0.0616734}, {3.29, 0.0615274}, {3.3, 0.0613816}, {3.31, 0.061236}, {3.32, \

0.0610905}, {3.33, 0.0609453}, {3.34, 0.0608003}, {3.35, 0.0606554}, {3.36, \
0.0605108}, {3.37, 0.0603663}, {3.38, 0.060222}, {3.39, 0.060078}, {3.4, \
0.0599341}, {3.41, 0.0597904}, {3.42, 0.0596469}, {3.43, 0.0595036}, {3.44, \
0.0593605}, {3.45, 0.0592176}, {3.46, 0.0590749}, {3.47, 0.0589324}, {3.48, \
0.0587901}, {3.49, 0.058648}, {3.5, 0.0585061}, {3.51, 0.0583644}, {3.52, \
0.0582229}, {3.53, 0.0580816}, {3.54, 0.0579405}, {3.55, 0.0577996}, {3.56, \
0.0576589}, {3.57, 0.0575184}, {3.58, 0.0573781}, {3.59, 0.057238}, {3.6, \
0.0570981}, {3.61, 0.0569584}, {3.62, 0.0568189}, {3.63, 0.0566796}, {3.64, \
0.0565405}, {3.65, 0.0564017}, {3.66, 0.056263}, {3.67, 0.0561246}, {3.68, \
0.0559863}, {3.69, 0.0558483}, {3.7, 0.0557104}, {3.71, 0.0555728}, {3.72, \
0.0554354}, {3.73, 0.0552981}, {3.74, 0.0551611}, {3.75, 0.0550243}, {3.76, \
0.0548877}, {3.77, 0.0547514}, {3.78, 0.0546152}, {3.79, 0.0544792}, {3.8, \
0.0543435}, {3.81, 0.0542079}, {3.82, 0.0540726}, {3.83, 0.0539375}, {3.84, \
0.0538026}, {3.85, 0.0536679}, {3.86, 0.0535334}, {3.87, 0.0533991}, {3.88, \
0.053265}, {3.89, 0.0531312}, {3.9, 0.0529975}, {3.91, 0.0528641}, {3.92, \
0.0527309}, {3.93, 0.0525979}, {3.94, 0.0524651}, {3.95, 0.0523326}, {3.96, \
0.0522002}, {3.97, 0.0520681}, {3.98, 0.0519362}, {3.99, 0.0518045}, {4.0, \
0.051673}, {4.01, 0.0515417}, {4.02, 0.0514107}, {4.03, 0.0512798}, {4.04, \
0.0511492}, {4.05, 0.0510188}, {4.06, 0.0508886}, {4.07, 0.0507586}, {4.08, \
0.0506289}, {4.09, 0.0504994}, {4.1, 0.0503701}, {4.11, 0.050241}, {4.12, \
0.0501121}, {4.13, 0.0499834}, {4.14, 0.049855}, {4.15, 0.0497268}, {4.16, \
0.0495988}, {4.17, 0.049471}, {4.18, 0.0493435}, {4.19, 0.0492162}, {4.2, \
0.0490891}, {4.21, 0.0489622}, {4.22, 0.0488355}, {4.23, 0.0487091}, {4.24, \
0.0485829}, {4.25, 0.0484569}, {4.26, 0.0483311}, {4.27, 0.0482056}, {4.28, \
0.0480802}, {4.29, 0.0479551}, {4.3, 0.0478303}, {4.31, 0.0477056}, {4.32, \
0.0475812}, {4.33, 0.047457}, {4.34, 0.047333}, {4.35, 0.0472093}, {4.36, \
0.0470858}, {4.37, 0.0469625}, {4.38, 0.0468394}, {4.39, 0.0467165}, {4.4, \
0.0465939}, {4.41, 0.0464715}, {4.42, 0.0463494}, {4.43, 0.0462275}, {4.44, \
0.0461057}, {4.45, 0.0459843}, {4.46, 0.045863}, {4.47, 0.045742}, {4.48, \
0.0456212}, {4.49, 0.0455007}, {4.5, 0.0453803}, {4.51, 0.0452602}, {4.52, \
0.0451403}, {4.53, 0.0450207}, {4.54, 0.0449013}, {4.55, 0.0447821}, {4.56, \
0.0446632}, {4.57, 0.0445444}, {4.58, 0.0444259}, {4.59, 0.0443077}, {4.6, \
0.0441897}, {4.61, 0.0440719}, {4.62, 0.0439543}, {4.63, 0.043837}, {4.64, \
0.0437199}, {4.65, 0.043603}, {4.66, 0.0434864}, {4.67, 0.04337}, {4.68, \
0.0432538}, {4.69, 0.0431379}, {4.7, 0.0430222}, {4.71, 0.0429067}, {4.72, \
0.0427915}, {4.73, 0.0426765}, {4.74, 0.0425617}, {4.75, 0.0424472}, {4.76, \
0.0423329}, {4.77, 0.0422188}, {4.78, 0.042105}, {4.79, 0.0419914}, {4.8, \
0.041878}, {4.81, 0.0417649}, {4.82, 0.041652}, {4.83, 0.0415393}, {4.84, \
0.0414269}, {4.85, 0.0413147}, {4.86, 0.0412028}, {4.87, 0.0410911}, {4.88, \
0.0409796}, {4.89, 0.0408684}, {4.9, 0.0407574}, {4.91, 0.0406466}, {4.92, \
0.0405361}, {4.93, 0.0404258}, {4.94, 0.0403157}, {4.95, 0.0402059}, {4.96, \
0.0400963}, {4.97, 0.039987}, {4.98, 0.0398778}, {4.99, 0.039769}, {5.0, \
0.0396603}

Profile of the Balmer-beta line for the observation parallel to the field

{0, 0.132826}, {0.01, 0.132825}, {0.02, 0.132822}, {0.03, 0.132816}, {0.04, \

0.132808}, {0.05, 0.132798}, {0.06, 0.132786}, {0.07, 0.132771}, {0.08, \
0.132755}, {0.09, 0.132736}, {0.1, 0.132714}, {0.11, 0.132691}, {0.12, \
0.132665}, {0.13, 0.132637}, {0.14, 0.132607}, {0.15, 0.132575}, {0.16, \
0.13254}, {0.17, 0.132504}, {0.18, 0.132465}, {0.19, 0.132423}, {0.2, \
0.13238}, {0.21, 0.132334}, {0.22, 0.132287}, {0.23, 0.132237}, {0.24, \
0.132185}, {0.25, 0.13213}, {0.26, 0.132074}, {0.27, 0.132015}, {0.28, \
0.131954}, {0.29, 0.131891}, {0.3, 0.131826}, {0.31, 0.131759}, {0.32, \
0.13169}, {0.33, 0.131618}, {0.34, 0.131544}, {0.35, 0.131469}, {0.36, \
0.131391}, {0.37, 0.131311}, {0.38, 0.131229}, {0.39, 0.131145}, {0.4, \
0.131058}, {0.41, 0.13097}, {0.42, 0.130879}, {0.43, 0.130787}, {0.44, \
0.130692}, {0.45, 0.130596}, {0.46, 0.130497}, {0.47, 0.130397}, {0.48, \
0.130294}, {0.49, 0.130189}, {0.5, 0.130083}, {0.51, 0.129974}, {0.52, \
0.129863}, {0.53, 0.129751}, {0.54, 0.129636}, {0.55, 0.12952}, {0.56, \
0.129401}, {0.57, 0.129281}, {0.58, 0.129159}, {0.59, 0.129034}, {0.6, \
0.128908}, {0.61, 0.12878}, {0.62, 0.12865}, {0.63, 0.128519}, {0.64, \
0.128385}, {0.65, 0.12825}, {0.66, 0.128112}, {0.67, 0.127973}, {0.68, \
0.127832}, {0.69, 0.12769}, {0.7, 0.127545}, {0.71, 0.127399}, {0.72, \
0.127251}, {0.73, 0.127101}, {0.74, 0.126949}, {0.75, 0.126796}, {0.76, \
0.126641}, {0.77, 0.126485}, {0.78, 0.126326}, {0.79, 0.126166}, {0.8, \
0.126005}, {0.81, 0.125841}, {0.82, 0.125676}, {0.83, 0.12551}, {0.84, \
0.125342}, {0.85, 0.125172}, {0.86, 0.125001}, {0.87, 0.124828}, {0.88, \
0.124653}, {0.89, 0.124477}, {0.9, 0.1243}, {0.91, 0.124121}, {0.92, \
0.12394}, {0.93, 0.123758}, {0.94, 0.123575}, {0.95, 0.12339}, {0.96, \
0.123203}, {0.97, 0.123015}, {0.98, 0.122826}, {0.99, 0.122635}, {1.0, \
0.122443}, {1.01, 0.12225}, {1.02, 0.122055}, {1.03, 0.121859}, {1.04, \
0.121662}, {1.05, 0.121463}, {1.06, 0.121263}, {1.07, 0.121062}, {1.08, \
0.120859}, {1.09, 0.120655}, {1.1, 0.12045}, {1.11, 0.120244}, {1.12, \
0.120036}, {1.13, 0.119828}, {1.14, 0.119618}, {1.15, 0.119407}, {1.16, \
0.119195}, {1.17, 0.118981}, {1.18, 0.118767}, {1.19, 0.118551}, {1.2, \
0.118335}, {1.21, 0.118117}, {1.22, 0.117898}, {1.23, 0.117678}, {1.24, \
0.117457}, {1.25, 0.117235}, {1.26, 0.117012}, {1.27, 0.116788}, {1.28, \
0.116563}, {1.29, 0.116338}, {1.3, 0.116111}, {1.31, 0.115883}, {1.32, \
0.115654}, {1.33, 0.115425}, {1.34, 0.115194}, {1.35, 0.114963}, {1.36, \
0.114731}, {1.37, 0.114498}, {1.38, 0.114264}, {1.39, 0.114029}, {1.4, \
0.113794}, {1.41, 0.113557}, {1.42, 0.11332}, {1.43, 0.113083}, {1.44, \
0.112844}, {1.45, 0.112605}, {1.46, 0.112365}, {1.47, 0.112124}, {1.48, \
0.111882}, {1.49, 0.11164}, {1.5, 0.111397}, {1.51, 0.111154}, {1.52, \
0.11091}, {1.53, 0.110665}, {1.54, 0.11042}, {1.55, 0.110174}, {1.56, \
0.109928}, {1.57, 0.10968}, {1.58, 0.109433}, {1.59, 0.109185}, {1.6, \
0.108936}, {1.61, 0.108687}, {1.62, 0.108437}, {1.63, 0.108187}, {1.64, \
0.107936}, {1.65, 0.107685}, {1.66, 0.107433}, {1.67, 0.107181}, {1.68, \
0.106928}, {1.69, 0.106675}, {1.7, 0.106422}, {1.71, 0.106168}, {1.72, \
0.105914}, {1.73, 0.10566}, {1.74, 0.105405}, {1.75, 0.105149}, {1.76, \
0.104894}, {1.77, 0.104638}, {1.78, 0.104382}, {1.79, 0.104125}, {1.8, \
0.103868}, {1.81, 0.103611}, {1.82, 0.103354}, {1.83, 0.103096}, {1.84, \

0.102838}, {1.85, 0.10258}, {1.86, 0.102322}, {1.87, 0.102063}, {1.88, \
0.101804}, {1.89, 0.101545}, {1.9, 0.101286}, {1.91, 0.101027}, {1.92, \
0.100767}, {1.93, 0.100508}, {1.94, 0.100248}, {1.95, 0.0999877}, {1.96, \
0.0997276}, {1.97, 0.0994674}, {1.98, 0.0992071}, {1.99, 0.0989466}, {2.0, \
0.0986861}, {2.01, 0.0984255}, {2.02, 0.0981648}, {2.03, 0.0979041}, {2.04, \
0.0976433}, {2.05, 0.0973825}, {2.06, 0.0971216}, {2.07, 0.0968607}, {2.08, \
0.0965998}, {2.09, 0.0963388}, {2.1, 0.0960779}, {2.11, 0.095817}, {2.12, \
0.095556}, {2.13, 0.0952951}, {2.14, 0.0950343}, {2.15, 0.0947734}, {2.16, \
0.0945126}, {2.17, 0.0942519}, {2.18, 0.0939912}, {2.19, 0.0937306}, {2.2, \
0.09347}, {2.21, 0.0932096}, {2.22, 0.0929492}, {2.23, 0.0926889}, {2.24, \
0.0924287}, {2.25, 0.0921687}, {2.26, 0.0919087}, {2.27, 0.0916489}, {2.28, \
0.0913892}, {2.29, 0.0911296}, {2.3, 0.0908701}, {2.31, 0.0906109}, {2.32, \
0.0903517}, {2.33, 0.0900927}, {2.34, 0.0898339}, {2.35, 0.0895753}, {2.36, \
0.0893168}, {2.37, 0.0890585}, {2.38, 0.0888004}, {2.39, 0.0885425}, {2.4, \
0.0882848}, {2.41, 0.0880273}, {2.42, 0.08777}, {2.43, 0.087513}, {2.44, \
0.0872561}, {2.45, 0.0869995}, {2.46, 0.0867431}, {2.47, 0.0864869}, {2.48, \
0.086231}, {2.49, 0.0859753}, {2.5, 0.0857199}, {2.51, 0.0854647}, {2.52, \
0.0852098}, {2.53, 0.0849551}, {2.54, 0.0847007}, {2.55, 0.0844466}, {2.56, \
0.0841927}, {2.57, 0.0839391}, {2.58, 0.0836858}, {2.59, 0.0834328}, {2.6, \
0.0831801}, {2.61, 0.0829277}, {2.62, 0.0826756}, {2.63, 0.0824237}, {2.64, \
0.0821722}, {2.65, 0.081921}, {2.66, 0.0816701}, {2.67, 0.0814195}, {2.68, \
0.0811692}, {2.69, 0.0809193}, {2.7, 0.0806697}, {2.71, 0.0804204}, {2.72, \
0.0801714}, {2.73, 0.0799228}, {2.74, 0.0796745}, {2.75, 0.0794265}, {2.76, \
0.0791789}, {2.77, 0.0789317}, {2.78, 0.0786848}, {2.79, 0.0784382}, {2.8, \
0.078192}, {2.81, 0.0779461}, {2.82, 0.0777006}, {2.83, 0.0774555}, {2.84, \
0.0772107}, {2.85, 0.0769663}, {2.86, 0.0767222}, {2.87, 0.0764786}, {2.88, \
0.0762353}, {2.89, 0.0759923}, {2.9, 0.0757498}, {2.91, 0.0755076}, {2.92, \
0.0752658}, {2.93, 0.0750244}, {2.94, 0.0747834}, {2.95, 0.0745427}, {2.96, \
0.0743025}, {2.97, 0.0740626}, {2.98, 0.0738231}, {2.99, 0.073584}, {3.0, \
0.0733453}, {3.01, 0.073107}, {3.02, 0.0728691}, {3.03, 0.0726316}, {3.04, \
0.0723946}, {3.05, 0.0721579}, {3.06, 0.0719216}, {3.07, 0.0716857}, {3.08, \
0.0714502}, {3.09, 0.0712151}, {3.1, 0.0709805}, {3.11, 0.0707462}, {3.12, \
0.0705124}, {3.13, 0.0702789}, {3.14, 0.0700459}, {3.15, 0.0698133}, {3.16, \
0.0695811}, {3.17, 0.0693494}, {3.18, 0.069118}, {3.19, 0.0688871}, {3.2, \
0.0686566}, {3.21, 0.0684265}, {3.22, 0.0681968}, {3.23, 0.0679675}, {3.24, \
0.0677387}, {3.25, 0.0675103}, {3.26, 0.0672824}, {3.27, 0.0670548}, {3.28, \
0.0668277}, {3.29, 0.066601}, {3.3, 0.0663747}, {3.31, 0.0661489}, {3.32, \
0.0659235}, {3.33, 0.0656985}, {3.34, 0.065474}, {3.35, 0.0652499}, {3.36, \
0.0650262}, {3.37, 0.064803}, {3.38, 0.0645802}, {3.39, 0.0643578}, {3.4, \
0.0641358}, {3.41, 0.0639143}, {3.42, 0.0636933}, {3.43, 0.0634727}, {3.44, \
0.0632525}, {3.45, 0.0630327}, {3.46, 0.0628134}, {3.47, 0.0625945}, {3.48, \
0.0623761}, {3.49, 0.0621581}, {3.5, 0.0619406}, {3.51, 0.0617235}, {3.52, \
0.0615068}, {3.53, 0.0612906}, {3.54, 0.0610748}, {3.55, 0.0608595}, {3.56, \
0.0606446}, {3.57, 0.0604301}, {3.58, 0.0602161}, {3.59, 0.0600025}, {3.6, \
0.0597894}, {3.61, 0.0595768}, {3.62, 0.0593645}, {3.63, 0.0591528}, {3.64, \

0.0589414}, {3.65, 0.0587305}, {3.66, 0.0585201}, {3.67, 0.0583101}, {3.68, \
0.0581006}, {3.69, 0.0578915}, {3.7, 0.0576828}, {3.71, 0.0574746}, {3.72, \
0.0572669}, {3.73, 0.0570596}, {3.74, 0.0568528}, {3.75, 0.0566464}, {3.76, \
0.0564404}, {3.77, 0.0562349}, {3.78, 0.0560299}, {3.79, 0.0558253}, {3.8, \
0.0556211}, {3.81, 0.0554175}, {3.82, 0.0552142}, {3.83, 0.0550114}, {3.84, \
0.0548091}, {3.85, 0.0546072}, {3.86, 0.0544058}, {3.87, 0.0542048}, {3.88, \
0.0540043}, {3.89, 0.0538042}, {3.9, 0.0536046}, {3.91, 0.0534055}, {3.92, \
0.0532068}, {3.93, 0.0530085}, {3.94, 0.0528107}, {3.95, 0.0526134}, {3.96, \
0.0524165}, {3.97, 0.0522201}, {3.98, 0.0520241}, {3.99, 0.0518286}, {4.0, \
0.0516336}, {4.01, 0.051439}, {4.02, 0.0512448}, {4.03, 0.0510511}, {4.04, \
0.0508579}, {4.05, 0.0506652}, {4.06, 0.0504728}, {4.07, 0.050281}, {4.08, \
0.0500896}, {4.09, 0.0498987}, {4.1, 0.0497082}, {4.11, 0.0495182}, {4.12, \
0.0493286}, {4.13, 0.0491395}, {4.14, 0.0489509}, {4.15, 0.0487627}, {4.16, \
0.048575}, {4.17, 0.0483877}, {4.18, 0.0482009}, {4.19, 0.0480145}, {4.2, \
0.0478287}, {4.21, 0.0476432}, {4.22, 0.0474583}, {4.23, 0.0472738}, {4.24, \
0.0470898}, {4.25, 0.0469062}, {4.26, 0.0467231}, {4.27, 0.0465404}, {4.28, \
0.0463582}, {4.29, 0.0461765}, {4.3, 0.0459952}, {4.31, 0.0458144}, {4.32, \
0.0456341}, {4.33, 0.0454542}, {4.34, 0.0452748}, {4.35, 0.0450958}, {4.36, \
0.0449173}, {4.37, 0.0447393}, {4.38, 0.0445618}, {4.39, 0.0443847}, {4.4, \
0.044208}, {4.41, 0.0440318}, {4.42, 0.0438561}, {4.43, 0.0436809}, {4.44, \
0.0435061}, {4.45, 0.0433318}, {4.46, 0.043158}, {4.47, 0.0429846}, {4.48, \
0.0428116}, {4.49, 0.0426392}, {4.5, 0.0424672}, {4.51, 0.0422957}, {4.52, \
0.0421246}, {4.53, 0.041954}, {4.54, 0.0417839}, {4.55, 0.0416142}, {4.56, \
0.041445}, {4.57, 0.0412763}, {4.58, 0.041108}, {4.59, 0.0409402}, {4.6, \
0.0407729}, {4.61, 0.040606}, {4.62, 0.0404396}, {4.63, 0.0402736}, {4.64, \
0.0401082}, {4.65, 0.0399432}, {4.66, 0.0397786}, {4.67, 0.0396145}, {4.68, \
0.0394509}, {4.69, 0.0392878}, {4.7, 0.0391251}, {4.71, 0.0389629}, {4.72, \
0.0388011}, {4.73, 0.0386399}, {4.74, 0.038479}, {4.75, 0.0383187}, {4.76, \
0.0381588}, {4.77, 0.0379994}, {4.78, 0.0378404}, {4.79, 0.037682}, {4.8, \
0.0375239}, {4.81, 0.0373664}, {4.82, 0.0372093}, {4.83, 0.0370527}, {4.84, \
0.0368965}, {4.85, 0.0367408}, {4.86, 0.0365856}, {4.87, 0.0364308}, {4.88, \
0.0362765}, {4.89, 0.0361227}, {4.9, 0.0359693}, {4.91, 0.0358164}, {4.92, \
0.035664}, {4.93, 0.035512}, {4.94, 0.0353605}, {4.95, 0.0352095}, {4.96, \
0.0350589}, {4.97, 0.0349088}, {4.98, 0.0347591}, {4.99, 0.0346099}, {5.0, \
0.0344612}

Profile of the Balmer-delta line for the observation perpendicular to the field

{0, 0.0415725}, {0.01, 0.0415724}, {0.02, 0.0415722}, {0.03, 0.0415719}, \
{0.04, 0.0415716}, {0.05, 0.041571}, {0.06, 0.0415704}, {0.07, 0.0415697}, \
{0.08, 0.0415688}, {0.09, 0.0415679}, {0.1, 0.0415668}, {0.11, 0.0415656}, \
{0.12, 0.0415643}, {0.13, 0.0415629}, {0.14, 0.0415614}, {0.15, 0.0415597}, \
{0.16, 0.041558}, {0.17, 0.0415561}, {0.18, 0.0415542}, {0.19, 0.0415521}, \
{0.2, 0.0415499}, {0.21, 0.0415475}, {0.22, 0.0415451}, {0.23, 0.0415426}, \
{0.24, 0.0415399}, {0.25, 0.0415372}, {0.26, 0.0415343}, {0.27, 0.0415313}, \
{0.28, 0.0415282}, {0.29, 0.041525}, {0.3, 0.0415216}, {0.31, 0.0415182}, \
{0.32, 0.0415146}, {0.33, 0.041511}, {0.34, 0.0415072}, {0.35, 0.0415033}, \

{0.36, 0.0414993}, {0.37, 0.0414952}, {0.38, 0.041491}, {0.39, 0.0414866}, \
{0.4, 0.0414822}, {0.41, 0.0414776}, {0.42, 0.041473}, {0.43, 0.0414682}, \
{0.44, 0.0414633}, {0.45, 0.0414583}, {0.46, 0.0414531}, {0.47, 0.0414479}, \
{0.48, 0.0414426}, {0.49, 0.0414371}, {0.5, 0.0414316}, {0.51, 0.0414259}, \
{0.52, 0.0414201}, {0.53, 0.0414142}, {0.54, 0.0414082}, {0.55, 0.0414021}, \
{0.56, 0.0413958}, {0.57, 0.0413895}, {0.58, 0.0413831}, {0.59, 0.0413765}, \
{0.6, 0.0413698}, {0.61, 0.041363}, {0.62, 0.0413561}, {0.63, 0.0413491}, \
{0.64, 0.041342}, {0.65, 0.0413348}, {0.66, 0.0413275}, {0.67, 0.04132}, \
{0.68, 0.0413125}, {0.69, 0.0413048}, {0.7, 0.0412971}, {0.71, 0.0412892}, \
{0.72, 0.0412812}, {0.73, 0.0412731}, {0.74, 0.0412649}, {0.75, 0.0412566}, \
{0.76, 0.0412482}, {0.77, 0.0412396}, {0.78, 0.041231}, {0.79, 0.0412222}, \
{0.8, 0.0412134}, {0.81, 0.0412044}, {0.82, 0.0411953}, {0.83, 0.0411862}, \
{0.84, 0.0411769}, {0.85, 0.0411675}, {0.86, 0.041158}, {0.87, 0.0411484}, \
{0.88, 0.0411386}, {0.89, 0.0411288}, {0.9, 0.0411189}, {0.91, 0.0411088}, \
{0.92, 0.0410987}, {0.93, 0.0410884}, {0.94, 0.0410781}, {0.95, 0.0410676}, \
{0.96, 0.0410571}, {0.97, 0.0410464}, {0.98, 0.0410356}, {0.99, 0.0410247}, \
{1.0, 0.0410137}, {1.01, 0.0410026}, {1.02, 0.0409914}, {1.03, 0.0409801}, \
{1.04, 0.0409687}, {1.05, 0.0409572}, {1.06, 0.0409455}, {1.07, 0.0409338}, \
{1.08, 0.040922}, {1.09, 0.04091}, {1.1, 0.040898}, {1.11, 0.0408859}, {1.12, \
0.0408736}, {1.13, 0.0408613}, {1.14, 0.0408488}, {1.15, 0.0408363}, {1.16, \
0.0408236}, {1.17, 0.0408108}, {1.18, 0.040798}, {1.19, 0.040785}, {1.2, \
0.0407719}, {1.21, 0.0407588}, {1.22, 0.0407455}, {1.23, 0.0407321}, {1.24, \
0.0407186}, {1.25, 0.040705}, {1.26, 0.0406914}, {1.27, 0.0406776}, {1.28, \
0.0406637}, {1.29, 0.0406497}, {1.3, 0.0406356}, {1.31, 0.0406215}, {1.32, \
0.0406072}, {1.33, 0.0405928}, {1.34, 0.0405783}, {1.35, 0.0405637}, {1.36, \
0.040549}, {1.37, 0.0405343}, {1.38, 0.0405194}, {1.39, 0.0405044}, {1.4, \
0.0404893}, {1.41, 0.0404742}, {1.42, 0.0404589}, {1.43, 0.0404435}, {1.44, \
0.040428}, {1.45, 0.0404125}, {1.46, 0.0403968}, {1.47, 0.040381}, {1.48, \
0.0403652}, {1.49, 0.0403492}, {1.5, 0.0403332}, {1.51, 0.040317}, {1.52, \
0.0403008}, {1.53, 0.0402845}, {1.54, 0.040268}, {1.55, 0.0402515}, {1.56, \
0.0402349}, {1.57, 0.0402181}, {1.58, 0.0402013}, {1.59, 0.0401844}, {1.6, \
0.0401674}, {1.61, 0.0401503}, {1.62, 0.0401331}, {1.63, 0.0401159}, {1.64, \
0.0400985}, {1.65, 0.040081}, {1.66, 0.0400634}, {1.67, 0.0400458}, {1.68, \
0.040028}, {1.69, 0.0400102}, {1.7, 0.0399923}, {1.71, 0.0399742}, {1.72, \
0.0399561}, {1.73, 0.0399379}, {1.74, 0.0399196}, {1.75, 0.0399012}, {1.76, \
0.0398827}, {1.77, 0.0398642}, {1.78, 0.0398455}, {1.79, 0.0398268}, {1.8, \
0.0398079}, {1.81, 0.039789}, {1.82, 0.03977}, {1.83, 0.0397509}, {1.84, \
0.0397317}, {1.85, 0.0397124}, {1.86, 0.039693}, {1.87, 0.0396736}, {1.88, \
0.039654}, {1.89, 0.0396344}, {1.9, 0.0396147}, {1.91, 0.0395949}, {1.92, \
0.039575}, {1.93, 0.039555}, {1.94, 0.0395349}, {1.95, 0.0395148}, {1.96, \
0.0394945}, {1.97, 0.0394742}, {1.98, 0.0394538}, {1.99, 0.0394333}, {2.0, \
0.0394127}, {2.01, 0.0393921}, {2.02, 0.0393713}, {2.03, 0.0393505}, {2.04, \
0.0393296}, {2.05, 0.0393086}, {2.06, 0.0392875}, {2.07, 0.0392663}, {2.08, \
0.0392451}, {2.09, 0.0392238}, {2.1, 0.0392023}, {2.11, 0.0391809}, {2.12, \
0.0391593}, {2.13, 0.0391376}, {2.14, 0.0391159}, {2.15, 0.0390941}, {2.16, \

0.0390722}, {2.17, 0.0390502}, {2.18, 0.0390282}, {2.19, 0.039006}, {2.2, \
0.0389838}, {2.21, 0.0389615}, {2.22, 0.0389392}, {2.23, 0.0389167}, {2.24, \
0.0388942}, {2.25, 0.0388716}, {2.26, 0.0388489}, {2.27, 0.0388262}, {2.28, \
0.0388033}, {2.29, 0.0387804}, {2.3, 0.0387574}, {2.31, 0.0387344}, {2.32, \
0.0387112}, {2.33, 0.038688}, {2.34, 0.0386647}, {2.35, 0.0386414}, {2.36, \
0.0386179}, {2.37, 0.0385944}, {2.38, 0.0385708}, {2.39, 0.0385472}, {2.4, \
0.0385234}, {2.41, 0.0384996}, {2.42, 0.0384757}, {2.43, 0.0384518}, {2.44, \
0.0384278}, {2.45, 0.0384037}, {2.46, 0.0383795}, {2.47, 0.0383553}, {2.48, \
0.0383309}, {2.49, 0.0383066}, {2.5, 0.0382821}, {2.51, 0.0382576}, {2.52, \
0.038233}, {2.53, 0.0382083}, {2.54, 0.0381836}, {2.55, 0.0381588}, {2.56, \
0.0381339}, {2.57, 0.038109}, {2.58, 0.038084}, {2.59, 0.0380589}, {2.6, \
0.0380338}, {2.61, 0.0380086}, {2.62, 0.0379833}, {2.63, 0.037958}, {2.64, \
0.0379326}, {2.65, 0.0379071}, {2.66, 0.0378816}, {2.67, 0.0378559}, {2.68, \
0.0378303}, {2.69, 0.0378045}, {2.7, 0.0377788}, {2.71, 0.0377529}, {2.72, \
0.037727}, {2.73, 0.037701}, {2.74, 0.0376749}, {2.75, 0.0376488}, {2.76, \
0.0376226}, {2.77, 0.0375964}, {2.78, 0.0375701}, {2.79, 0.0375437}, {2.8, \
0.0375173}, {2.81, 0.0374908}, {2.82, 0.0374643}, {2.83, 0.0374377}, {2.84, \
0.037411}, {2.85, 0.0373843}, {2.86, 0.0373575}, {2.87, 0.0373306}, {2.88, \
0.0373037}, {2.89, 0.0372768}, {2.9, 0.0372498}, {2.91, 0.0372227}, {2.92, \
0.0371955}, {2.93, 0.0371684}, {2.94, 0.0371411}, {2.95, 0.0371138}, {2.96, \
0.0370864}, {2.97, 0.037059}, {2.98, 0.0370315}, {2.99, 0.037004}, {3.0, \
0.0369764}, {3.01, 0.0369488}, {3.02, 0.0369211}, {3.03, 0.0368933}, {3.04, \
0.0368655}, {3.05, 0.0368377}, {3.06, 0.0368098}, {3.07, 0.0367818}, {3.08, \
0.0367538}, {3.09, 0.0367257}, {3.1, 0.0366976}, {3.11, 0.0366694}, {3.12, \
0.0366412}, {3.13, 0.0366129}, {3.14, 0.0365846}, {3.15, 0.0365562}, {3.16, \
0.0365278}, {3.17, 0.0364993}, {3.18, 0.0364708}, {3.19, 0.0364423}, {3.2, \
0.0364136}, {3.21, 0.036385}, {3.22, 0.0363562}, {3.23, 0.0363275}, {3.24, \
0.0362987}, {3.25, 0.0362698}, {3.26, 0.0362409}, {3.27, 0.0362119}, {3.28, \
0.0361829}, {3.29, 0.0361539}, {3.3, 0.0361248}, {3.31, 0.0360957}, {3.32, \
0.0360665}, {3.33, 0.0360373}, {3.34, 0.036008}, {3.35, 0.0359787}, {3.36, \
0.0359493}, {3.37, 0.0359199}, {3.38, 0.0358905}, {3.39, 0.035861}, {3.4, \
0.0358315}, {3.41, 0.0358019}, {3.42, 0.0357723}, {3.43, 0.0357426}, {3.44, \
0.0357129}, {3.45, 0.0356832}, {3.46, 0.0356534}, {3.47, 0.0356236}, {3.48, \
0.0355937}, {3.49, 0.0355638}, {3.5, 0.0355339}, {3.51, 0.0355039}, {3.52, \
0.0354739}, {3.53, 0.0354439}, {3.54, 0.0354138}, {3.55, 0.0353837}, {3.56, \
0.0353535}, {3.57, 0.0353233}, {3.58, 0.035293}, {3.59, 0.0352628}, {3.6, \
0.0352325}, {3.61, 0.0352021}, {3.62, 0.0351717}, {3.63, 0.0351413}, {3.64, \
0.0351109}, {3.65, 0.0350804}, {3.66, 0.0350498}, {3.67, 0.0350193}, {3.68, \
0.0349887}, {3.69, 0.0349581}, {3.7, 0.0349274}, {3.71, 0.0348967}, {3.72, \
0.034866}, {3.73, 0.0348353}, {3.74, 0.0348045}, {3.75, 0.0347737}, {3.76, \
0.0347428}, {3.77, 0.0347119}, {3.78, 0.034681}, {3.79, 0.0346501}, {3.8, \
0.0346191}, {3.81, 0.0345881}, {3.82, 0.0345571}, {3.83, 0.034526}, {3.84, \
0.034495}, {3.85, 0.0344638}, {3.86, 0.0344327}, {3.87, 0.0344015}, {3.88, \
0.0343703}, {3.89, 0.0343391}, {3.9, 0.0343079}, {3.91, 0.0342766}, {3.92, \
0.0342453}, {3.93, 0.034214}, {3.94, 0.0341826}, {3.95, 0.0341512}, {3.96, \

0.0341198}, {3.97, 0.0340884}, {3.98, 0.0340569}, {3.99, 0.0340255}, {4.0, \
0.033994}, {4.01, 0.0339624}, {4.02, 0.0339309}, {4.03, 0.0338993}, {4.04, \
0.0338677}, {4.05, 0.0338361}, {4.06, 0.0338045}, {4.07, 0.0337728}, {4.08, \
0.0337411}, {4.09, 0.0337094}, {4.1, 0.0336777}, {4.11, 0.0336459}, {4.12, \
0.0336142}, {4.13, 0.0335824}, {4.14, 0.0335506}, {4.15, 0.0335188}, {4.16, \
0.0334869}, {4.17, 0.0334551}, {4.18, 0.0334232}, {4.19, 0.0333913}, {4.2, \
0.0333594}, {4.21, 0.0333274}, {4.22, 0.0332955}, {4.23, 0.0332635}, {4.24, \
0.0332315}, {4.25, 0.0331995}, {4.26, 0.0331675}, {4.27, 0.0331355}, {4.28, \
0.0331035}, {4.29, 0.0330714}, {4.3, 0.0330393}, {4.31, 0.0330072}, {4.32, \
0.0329751}, {4.33, 0.032943}, {4.34, 0.0329109}, {4.35, 0.0328787}, {4.36, \
0.0328465}, {4.37, 0.0328144}, {4.38, 0.0327822}, {4.39, 0.03275}, {4.4, \
0.0327178}, {4.41, 0.0326856}, {4.42, 0.0326533}, {4.43, 0.0326211}, {4.44, \
0.0325888}, {4.45, 0.0325565}, {4.46, 0.0325243}, {4.47, 0.032492}, {4.48, \
0.0324597}, {4.49, 0.0324274}, {4.5, 0.0323951}, {4.51, 0.0323627}, {4.52, \
0.0323304}, {4.53, 0.032298}, {4.54, 0.0322657}, {4.55, 0.0322333}, {4.56, \
0.032201}, {4.57, 0.0321686}, {4.58, 0.0321362}, {4.59, 0.0321038}, {4.6, \
0.0320714}, {4.61, 0.032039}, {4.62, 0.0320066}, {4.63, 0.0319742}, {4.64, \
0.0319418}, {4.65, 0.0319094}, {4.66, 0.0318769}, {4.67, 0.0318445}, {4.68, \
0.031812}, {4.69, 0.0317796}, {4.7, 0.0317472}, {4.71, 0.0317147}, {4.72, \
0.0316823}, {4.73, 0.0316498}, {4.74, 0.0316173}, {4.75, 0.0315849}, {4.76, \
0.0315524}, {4.77, 0.0315199}, {4.78, 0.0314875}, {4.79, 0.031455}, {4.8, \
0.0314225}, {4.81, 0.03139}, {4.82, 0.0313576}, {4.83, 0.0313251}, {4.84, \
0.0312926}, {4.85, 0.0312601}, {4.86, 0.0312277}, {4.87, 0.0311952}, {4.88, \
0.0311627}, {4.89, 0.0311302}, {4.9, 0.0310978}, {4.91, 0.0310653}, {4.92, \
0.0310328}, {4.93, 0.0310003}, {4.94, 0.0309679}, {4.95, 0.0309354}, {4.96, \
0.0309029}, {4.97, 0.0308705}, {4.98, 0.030838}, {4.99, 0.0308056}, {5.0, \
0.0307731}

Profile of the Balmer-delta line for the observation parallel to the field

{0, 0.0567506}, {0.01, 0.0567505}, {0.02, 0.0567502}, {0.03, 0.0567497}, \
{0.04, 0.0567489}, {0.05, 0.056748}, {0.06, 0.0567468}, {0.07, 0.0567455}, \
{0.08, 0.0567439}, {0.09, 0.0567421}, {0.1, 0.0567401}, {0.11, 0.0567379}, \
{0.12, 0.0567354}, {0.13, 0.0567328}, {0.14, 0.0567299}, {0.15, 0.0567269}, \
{0.16, 0.0567236}, {0.17, 0.0567201}, {0.18, 0.0567164}, {0.19, 0.0567125}, \
{0.2, 0.0567084}, {0.21, 0.0567041}, {0.22, 0.0566995}, {0.23, 0.0566948}, \
{0.24, 0.0566898}, {0.25, 0.0566847}, {0.26, 0.0566793}, {0.27, 0.0566737}, \
{0.28, 0.0566679}, {0.29, 0.0566619}, {0.3, 0.0566557}, {0.31, 0.0566493}, \
{0.32, 0.0566426}, {0.33, 0.0566358}, {0.34, 0.0566287}, {0.35, 0.0566215}, \
{0.36, 0.056614}, {0.37, 0.0566063}, {0.38, 0.0565984}, {0.39, 0.0565903}, \
{0.4, 0.056582}, {0.41, 0.0565735}, {0.42, 0.0565648}, {0.43, 0.0565558}, \
{0.44, 0.0565467}, {0.45, 0.0565373}, {0.46, 0.0565278}, {0.47, 0.056518}, \
{0.48, 0.056508}, {0.49, 0.0564978}, {0.5, 0.0564874}, {0.51, 0.0564768}, \
{0.52, 0.056466}, {0.53, 0.056455}, {0.54, 0.0564438}, {0.55, 0.0564324}, \
{0.56, 0.0564207}, {0.57, 0.0564089}, {0.58, 0.0563969}, {0.59, 0.0563846}, \
{0.6, 0.0563722}, {0.61, 0.0563595}, {0.62, 0.0563466}, {0.63, 0.0563336}, \
{0.64, 0.0563203}, {0.65, 0.0563068}, {0.66, 0.0562931}, {0.67, 0.0562792}, \

{0.68, 0.0562651}, {0.69, 0.0562508}, {0.7, 0.0562363}, {0.71, 0.0562216}, \
{0.72, 0.0562067}, {0.73, 0.0561916}, {0.74, 0.0561763}, {0.75, 0.0561608}, \
{0.76, 0.056145}, {0.77, 0.0561291}, {0.78, 0.056113}, {0.79, 0.0560967}, \
{0.8, 0.0560801}, {0.81, 0.0560634}, {0.82, 0.0560465}, {0.83, 0.0560293}, \
{0.84, 0.056012}, {0.85, 0.0559945}, {0.86, 0.0559768}, {0.87, 0.0559588}, \
{0.88, 0.0559407}, {0.89, 0.0559224}, {0.9, 0.0559038}, {0.91, 0.0558851}, \
{0.92, 0.0558662}, {0.93, 0.0558471}, {0.94, 0.0558277}, {0.95, 0.0558082}, \
{0.96, 0.0557885}, {0.97, 0.0557686}, {0.98, 0.0557485}, {0.99, 0.0557282}, \
{1.0, 0.0557077}, {1.01, 0.055687}, {1.02, 0.0556661}, {1.03, 0.055645}, \
{1.04, 0.0556237}, {1.05, 0.0556023}, {1.06, 0.0555806}, {1.07, 0.0555587}, \
{1.08, 0.0555367}, {1.09, 0.0555144}, {1.1, 0.055492}, {1.11, 0.0554693}, \
{1.12, 0.0554465}, {1.13, 0.0554235}, {1.14, 0.0554003}, {1.15, 0.0553769}, \
{1.16, 0.0553533}, {1.17, 0.0553295}, {1.18, 0.0553055}, {1.19, 0.0552814}, \
{1.2, 0.055257}, {1.21, 0.0552325}, {1.22, 0.0552077}, {1.23, 0.0551828}, \
{1.24, 0.0551577}, {1.25, 0.0551324}, {1.26, 0.0551069}, {1.27, 0.0550813}, \
{1.28, 0.0550554}, {1.29, 0.0550294}, {1.3, 0.0550032}, {1.31, 0.0549767}, \
{1.32, 0.0549501}, {1.33, 0.0549234}, {1.34, 0.0548964}, {1.35, 0.0548692}, \
{1.36, 0.0548419}, {1.37, 0.0548144}, {1.38, 0.0547867}, {1.39, 0.0547588}, \
{1.4, 0.0547308}, {1.41, 0.0547025}, {1.42, 0.0546741}, {1.43, 0.0546455}, \
{1.44, 0.0546167}, {1.45, 0.0545877}, {1.46, 0.0545586}, {1.47, 0.0545293}, \
{1.48, 0.0544998}, {1.49, 0.0544701}, {1.5, 0.0544402}, {1.51, 0.0544102}, \
{1.52, 0.05438}, {1.53, 0.0543496}, {1.54, 0.054319}, {1.55, 0.0542883}, \
{1.56, 0.0542574}, {1.57, 0.0542263}, {1.58, 0.054195}, {1.59, 0.0541636}, \
{1.6, 0.054132}, {1.61, 0.0541002}, {1.62, 0.0540683}, {1.63, 0.0540361}, \
{1.64, 0.0540038}, {1.65, 0.0539714}, {1.66, 0.0539387}, {1.67, 0.0539059}, \
{1.68, 0.0538729}, {1.69, 0.0538398}, {1.7, 0.0538065}, {1.71, 0.053773}, \
{1.72, 0.0537394}, {1.73, 0.0537055}, {1.74, 0.0536716}, {1.75, 0.0536374}, \
{1.76, 0.0536031}, {1.77, 0.0535686}, {1.78, 0.053534}, {1.79, 0.0534991}, \
{1.8, 0.0534642}, {1.81, 0.053429}, {1.82, 0.0533937}, {1.83, 0.0533583}, \
{1.84, 0.0533226}, {1.85, 0.0532869}, {1.86, 0.0532509}, {1.87, 0.0532148}, \
{1.88, 0.0531785}, {1.89, 0.0531421}, {1.9, 0.0531055}, {1.91, 0.0530688}, \
{1.92, 0.0530319}, {1.93, 0.0529948}, {1.94, 0.0529576}, {1.95, 0.0529202}, \
{1.96, 0.0528827}, {1.97, 0.052845}, {1.98, 0.0528072}, {1.99, 0.0527692}, \
{2.0, 0.0527311}, {2.01, 0.0526928}, {2.02, 0.0526543}, {2.03, 0.0526157}, \
{2.04, 0.052577}, {2.05, 0.0525381}, {2.06, 0.052499}, {2.07, 0.0524598}, \
{2.08, 0.0524205}, {2.09, 0.052381}, {2.1, 0.0523413}, {2.11, 0.0523015}, \
{2.12, 0.0522616}, {2.13, 0.0522215}, {2.14, 0.0521812}, {2.15, 0.0521409}, \
{2.16, 0.0521003}, {2.17, 0.0520597}, {2.18, 0.0520188}, {2.19, 0.0519779}, \
{2.2, 0.0519368}, {2.21, 0.0518955}, {2.22, 0.0518542}, {2.23, 0.0518126}, \
{2.24, 0.051771}, {2.25, 0.0517292}, {2.26, 0.0516872}, {2.27, 0.0516451}, \
{2.28, 0.0516029}, {2.29, 0.0515606}, {2.3, 0.0515181}, {2.31, 0.0514754}, \
{2.32, 0.0514327}, {2.33, 0.0513898}, {2.34, 0.0513467}, {2.35, 0.0513035}, \
{2.36, 0.0512602}, {2.37, 0.0512168}, {2.38, 0.0511732}, {2.39, 0.0511295}, \
{2.4, 0.0510857}, {2.41, 0.0510417}, {2.42, 0.0509976}, {2.43, 0.0509534}, \
{2.44, 0.0509091}, {2.45, 0.0508646}, {2.46, 0.05082}, {2.47, 0.0507752}, \

{2.48, 0.0507304}, {2.49, 0.0506854}, {2.5, 0.0506403}, {2.51, 0.050595}, \
{2.52, 0.0505497}, {2.53, 0.0505042}, {2.54, 0.0504586}, {2.55, 0.0504129}, \
{2.56, 0.050367}, {2.57, 0.050321}, {2.58, 0.0502749}, {2.59, 0.0502287}, \
{2.6, 0.0501824}, {2.61, 0.050136}, {2.62, 0.0500894}, {2.63, 0.0500427}, \
{2.64, 0.0499959}, {2.65, 0.049949}, {2.66, 0.0499019}, {2.67, 0.0498548}, \
{2.68, 0.0498075}, {2.69, 0.0497602}, {2.7, 0.0497127}, {2.71, 0.0496651}, \
{2.72, 0.0496173}, {2.73, 0.0495695}, {2.74, 0.0495216}, {2.75, 0.0494735}, \
{2.76, 0.0494254}, {2.77, 0.0493771}, {2.78, 0.0493287}, {2.79, 0.0492803}, \
{2.8, 0.0492317}, {2.81, 0.049183}, {2.82, 0.0491342}, {2.83, 0.0490853}, \
{2.84, 0.0490363}, {2.85, 0.0489871}, {2.86, 0.0489379}, {2.87, 0.0488886}, \
{2.88, 0.0488392}, {2.89, 0.0487897}, {2.9, 0.04874}, {2.91, 0.0486903}, \
{2.92, 0.0486405}, {2.93, 0.0485906}, {2.94, 0.0485406}, {2.95, 0.0484904}, \
{2.96, 0.0484402}, {2.97, 0.0483899}, {2.98, 0.0483395}, {2.99, 0.048289}, \
{3.0, 0.0482384}, {3.01, 0.0481877}, {3.02, 0.0481369}, {3.03, 0.0480861}, \
{3.04, 0.0480351}, {3.05, 0.047984}, {3.06, 0.0479329}, {3.07, 0.0478816}, \
{3.08, 0.0478303}, {3.09, 0.0477789}, {3.1, 0.0477274}, {3.11, 0.0476758}, \
{3.12, 0.0476241}, {3.13, 0.0475723}, {3.14, 0.0475205}, {3.15, 0.0474686}, \
{3.16, 0.0474165}, {3.17, 0.0473644}, {3.18, 0.0473122}, {3.19, 0.04726}, \
{3.2, 0.0472076}, {3.21, 0.0471552}, {3.22, 0.0471027}, {3.23, 0.0470501}, \
{3.24, 0.0469974}, {3.25, 0.0469446}, {3.26, 0.0468918}, {3.27, 0.0468389}, \
{3.28, 0.0467859}, {3.29, 0.0467329}, {3.3, 0.0466797}, {3.31, 0.0466265}, \
{3.32, 0.0465732}, {3.33, 0.0465199}, {3.34, 0.0464664}, {3.35, 0.0464129}, \
{3.36, 0.0463593}, {3.37, 0.0463057}, {3.38, 0.046252}, {3.39, 0.0461982}, \
{3.4, 0.0461443}, {3.41, 0.0460904}, {3.42, 0.0460364}, {3.43, 0.0459824}, \
{3.44, 0.0459282}, {3.45, 0.045874}, {3.46, 0.0458198}, {3.47, 0.0457654}, \
{3.48, 0.0457111}, {3.49, 0.0456566}, {3.5, 0.0456021}, {3.51, 0.0455475}, \
{3.52, 0.0454929}, {3.53, 0.0454382}, {3.54, 0.0453834}, {3.55, 0.0453286}, \
{3.56, 0.0452737}, {3.57, 0.0452188}, {3.58, 0.0451638}, {3.59, 0.0451087}, \
{3.6, 0.0450536}, {3.61, 0.0449985}, {3.62, 0.0449432}, {3.63, 0.044888}, \
{3.64, 0.0448326}, {3.65, 0.0447772}, {3.66, 0.0447218}, {3.67, 0.0446663}, \
{3.68, 0.0446108}, {3.69, 0.0445552}, {3.7, 0.0444995}, {3.71, 0.0444439}, \
{3.72, 0.0443881}, {3.73, 0.0443323}, {3.74, 0.0442765}, {3.75, 0.0442206}, \
{3.76, 0.0441647}, {3.77, 0.0441087}, {3.78, 0.0440527}, {3.79, 0.0439966}, \
{3.8, 0.0439405}, {3.81, 0.0438843}, {3.82, 0.0438281}, {3.83, 0.0437719}, \
{3.84, 0.0437156}, {3.85, 0.0436593}, {3.86, 0.0436029}, {3.87, 0.0435465}, \
{3.88, 0.0434901}, {3.89, 0.0434336}, {3.9, 0.043377}, {3.91, 0.0433205}, \
{3.92, 0.0432639}, {3.93, 0.0432073}, {3.94, 0.0431506}, {3.95, 0.0430939}, \
{3.96, 0.0430371}, {3.97, 0.0429804}, {3.98, 0.0429236}, {3.99, 0.0428667}, \
{4.0, 0.0428098}, {4.01, 0.0427529}, {4.02, 0.042696}, {4.03, 0.042639}, \
{4.04, 0.042582}, {4.05, 0.042525}, {4.06, 0.042468}, {4.07, 0.0424109}, \
{4.08, 0.0423538}, {4.09, 0.0422966}, {4.1, 0.0422395}, {4.11, 0.0421823}, \
{4.12, 0.0421251}, {4.13, 0.0420678}, {4.14, 0.0420106}, {4.15, 0.0419533}, \
{4.16, 0.041896}, {4.17, 0.0418386}, {4.18, 0.0417813}, {4.19, 0.0417239}, \
{4.2, 0.0416665}, {4.21, 0.0416091}, {4.22, 0.0415517}, {4.23, 0.0414942}, \
{4.24, 0.0414367}, {4.25, 0.0413793}, {4.26, 0.0413217}, {4.27, 0.0412642}, \

```
{4.28, 0.0412067}, {4.29, 0.0411491}, {4.3, 0.0410916}, {4.31, 0.041034}, \
{4.32, 0.0409764}, {4.33, 0.0409188}, {4.34, 0.0408611}, {4.35, 0.0408035}, \
{4.36, 0.0407459}, {4.37, 0.0406882}, {4.38, 0.0406305}, {4.39, 0.0405729}, \
{4.4, 0.0405152}, {4.41, 0.0404575}, {4.42, 0.0403998}, {4.43, 0.040342}, \
{4.44, 0.0402843}, {4.45, 0.0402266}, {4.46, 0.0401689}, {4.47, 0.0401111}, \
{4.48, 0.0400534}, {4.49, 0.0399956}, {4.5, 0.0399379}, {4.51, 0.0398801}, \
{4.52, 0.0398224}, {4.53, 0.0397646}, {4.54, 0.0397068}, {4.55, 0.0396491}, \
{4.56, 0.0395913}, {4.57, 0.0395336}, {4.58, 0.0394758}, {4.59, 0.039418}, \
{4.6, 0.0393603}, {4.61, 0.0393025}, {4.62, 0.0392447}, {4.63, 0.039187}, \
{4.64, 0.0391292}, {4.65, 0.0390715}, {4.66, 0.0390137}, {4.67, 0.038956}, \
{4.68, 0.0388983}, {4.69, 0.0388406}, {4.7, 0.0387828}, {4.71, 0.0387251}, \
{4.72, 0.0386674}, {4.73, 0.0386097}, {4.74, 0.038552}, {4.75, 0.0384943}, \
{4.76, 0.0384367}, {4.77, 0.038379}, {4.78, 0.0383214}, {4.79, 0.0382637}, \
{4.8, 0.0382061}, {4.81, 0.0381485}, {4.82, 0.0380909}, {4.83, 0.0380333}, \
{4.84, 0.0379757}, {4.85, 0.0379181}, {4.86, 0.0378606}, {4.87, 0.0378031}, \
{4.88, 0.0377455}, {4.89, 0.037688}, {4.9, 0.0376305}, {4.91, 0.0375731}, \
{4.92, 0.0375156}, {4.93, 0.0374582}, {4.94, 0.0374008}, {4.95, 0.0373434}, \
{4.96, 0.037286}, {4.97, 0.0372286}, {4.98, 0.0371713}, {4.99, 0.037114}, \
{5.0, 0.0370567}
```

IOP Publishing

Simple Atomic and Molecular Systems
New results and applications
Eugene Oks

Appendix D

Brief description of the application of Dirac's generalized Hamiltonian dynamics (GHD) to microscopic systems of particles

The central point of Dirac's generalized Hamiltonian dynamics (GHD) is the introduction of a *classical* generalized Hamilton function

$$H_g(q, p) = H(q, p) + u_m\phi_m(q, p), \qquad m = 1, 2, \dots, M, \tag{D.1}$$

where $H(q, p)$ is the conventional classical Hamilton function and the second term is a linear combination of constraints [1–3]. Here and below, summation over a twice repeated suffix is understood. The coefficients u_m can be determined from so-called consistency conditions

$$[\phi_{m'}, H] + u_m[\phi_{m'}, \phi_m] \cong 0. \qquad (m' = 1, 2, \dots, M), \tag{D.2}$$

where $[f, g]$ is the usual Poisson bracket.

Oks and Uzer [4] further developed Dirac's GHD using classical integrals of the motion as constraints—see also [5, 6]. Specifically, for hydrogenic atoms/ions they used components of the angular momentum and of the Runge–Lenz vector as the constraints. This turned out to be sufficient for proving the existence of classical non-radiating states.

To obtain further results, Oks and Uzer [4] employed Plank's hypothesis (later incorporated by Einstein into his photoelectric law—see e.g. the textbook [7]), which can be formulated as follows. In the process of radiation the change of the energy between the initial and final states ΔE should be proportional to the average frequency of the radiation $\langle\omega\rangle$ in this process. As a result, Oks and Uzer [4] obtained the expression for the infinite set of classical non-radiating states E_k (where $k = 1, 2, 3, \dots, \infty$) coinciding with the corresponding set of quantum stationary states.

Later Dirac's GHD was successfully applied to other microscopic systems and to finding classical non-radiating states in these systems. Examples are pairs of particles

doi:10.1088/978-0-7503-3679-6ch14

interacting through a modified Coulomb potential [8] and a spherical harmonic oscillator [9].

References

[1] Dirac P A M 1950 *Can. J. Math.* **2** 129–48
[2] Dirac P A M 1958 *Proc. R. Soc.* A **246** 326–32
[3] Dirac P A M 1964 *Lectures on Quantum Mechanics* (New York: Academic) reprinted by Dover 2001
[4] Oks E and Uzer T 2002 *J. Phys. B: At. Mol. Opt. Phys.* **35** 165–73
[5] Oks E 2015 *Breaking Paradigms in Atomic and Molecular Physics* (Singapore: World Scientific) ch 4
[6] Oks E 2019 *Unexpected Similarities of the Universe with Atomic and Molecular Systems: What a Beautiful World* (Bristol: IOP Publishing) appendix A
[7] Loeb L L 1938 *Atomic Structure* (London: Wiley) ch VI.2
[8] Camarena A and Oks E 2010 *Int. Rev. At. Mol. Phys.* **1** 143–60
[9] Oks E 2020 *Symmetry* **12** 1130

Lightning Source UK Ltd.
Milton Keynes UK
UKHW051905030321
379702UK00003B/41